Assessment in transition

Assessment in transition

learning, monitoring and selection in international perspective

Edited by
Angela Little and Alison Wolf

Pergamon

U.K.	Elsevier Science Ltd, The Boulevard, Langford Lane, Kidlington, Oxford OX5 1GB, U.K.
U.S.A.	Elsevier Science Inc., 660 White Plains Road, Tarrytown, New York 10591-5153, U.S.A.
JAPAN	Elsevier Science Japan, Higashi Azabu 1-chome Building 4F, 1-9-15, Higashi Azabu, Minato-ku, Tokyo 106, Japan

First edition 1996

Library of Congress Cataloging in Publication Data

Assessment in transition: learning, monitoring and selection in international perspective/edited by Angela Little & Alison Wolf.
—1st ed.
p. cm
Includes indexes.
1. Education tests and measurements—Cross-cultural studies.
2. Examinations—Cross-cultural studies.
I. Little, Angela.
II. Wolf, Alison.
LB3051.L372 1996
371.2'6–dc20

British Library Cataloguing in Publication Data

A catalogue record for this book is available from the British Library.

ISBN 0 08 042767 7

Printed in Great Britain by Biddles Ltd, Guildford and King's Lynn

Contents

*Part Three   Country Case Studies: The Backwash of Selection—
Contrasting Contexts, Common Dilemmas*

Part Four   Resolving the Tensions? Possibilities and Limits

Acknowledgements

In July 1993 30 educational assessment researchers and practitioners met at the Institute of Education, University of London, to discuss the theme of this book. The present volume grew out of that meeting; and we are grateful to all for their insights, to those who travelled thousands of miles to join us and to those who would contribute chapters subsequently. Financial support for that meeting was generously afforded by the British Council, the World Bank, UNESCO and participants' respective institutions as well as by personal resources.

We are especially grateful to Magdalen Meade for technical support, including responsibility for diagrams and tables, and John Lowe for able editorial support. We would also like to thank Fernando Marhuendo and Roger Flavell for assistance and comments, and the Institute of Education more generally for affording the environment in which to reflect and write.

Angela Little
Alison Wolf
International Centre for Research on Assessment
Institute of Education
University of London

Foreword

In 1990 the World Conference on Education for All (WCEFA) in Jomtien, Thailand affirmed a global commitment to the achievement of education for all young persons. This commitment went beyond those of earlier declarations by emphasizing not merely access to education—larger enrolments and increased participation rates—but the achievement of valued outcomes. The 1990 declaration argues that:

> whether or not expanded educational opportunities will translate into meaningful development—for an individual or society—depends ultimately on whether people actually learn as a result of those opportunities: i.e. whether they incorporate useful knowledge, reasoning ability, skills and values. (*World Declaration on Education for All*, 1990, Article 4)

The focus is therefore on learning and learning acquisition—but also, by necessity, on *assessment*. For only by measuring and assessing can we tell whether and what people have learned. A move from access and quantity of education to outcomes and quality translates into a growing emphasis on assessment. This emphasis is now evident worldwide, in developing and developed countries alike.

The power of assessment to drive the curriculum is by now well attested. It is, indeed, one of the few things on which everyone in the world of education would agree. However, the link between assessment and curriculum is not as straightforward as many believe, nor simple to establish. On the contrary, there are many tensions inherent in educational assessment, and current reforms have made these increasingly clear.

The emphasis on measuring learning outcomes reflects the desire to promote learning for all: but also the demands of funding agencies—national or international—for greater accountability and cost-effectiveness. As education and training absorb ever greater sums, in both absolute and relative terms, such demands also grow. Governments, in particular, want to monitor, and also to control education, in order to ensure 'value for money'. At first sight, this objective, and that of promoting learning, may seem closely linked, since both are concerned with individuals' achievement. In practice, as this book makes clear, tensions between the two quickly become evident.

Moreover, the new emphasis on assessment as a tool to promote 'quality' across a whole educational system creates additional demands while leaving the old untouched. The traditional requirement of assessment was that it should provide an efficient way to rank and select among individuals. Far from disappearing, assessment for selection is actually of growing importance. The goals of monitoring and learning exist alongside it, but in no sense replace it.

This book addresses the growing tensions between the goals of contemporary educational assessment. For our purposes, the term 'assessment' is used broadly, to cover all judgements of educational performance which are used in individual or aggregated form, for one or more purposes and by a range of persons and institutions: a definition which reflects the growing penetration of economic and social life by such judgements. The focus is, throughout, comparative, historical and international. The new demands on assessment are, as we have emphasized, worldwide: but different contexts and different policies yield different responses to apparently similar demands. Such knowledge is invaluable in clarifying trends and possibilities.

The roles of assessment: comparative, historical and international variations

The dominant role of educational assessment throughout much of history has been educational and occupational selection. The Chinese Imperial civil service examinations were established over 2000 years ago and created the precedent for occupational entry on the basis of performance in public examinations.

Several chapters in this book attest to the continuing dominance of assessment for selection purposes. The fragmented nature of selection systems in South Africa under apartheid and the new and challenging role for selective assessment as the country moves towards an integrated system of education and training are described by Peliwe Lolwana (Chapter 10). Chan-Jong Kim (Chapter 12) discusses how many Koreans believe entrance examinations to subsequent levels of education to be a source of major educational problems, while Sunderi Kariyewasam (Chapter 13) describes the interest of Sri Lankans in examinations as 'obsessive'.

The effects of selective assessment will be determined in part by the selection ratio. Tony Somerset (Chapter 16) describes the contrasting cases of Nepal and Tanzania where primary leaving certificates govern access to secondary school places. In Nepal in the 1980s the transition rate to secondary school was around 90%, and in Tanzania only 4%. In Nepal the passing of the selection examination 'is a formality rather than a real challenge', whereas in Tanzania 'the secondary school intake was so small that those who completed the course four years later could be almost certain that a higher secondary place, formal job, or other desirable opportunity would be open to them'.

A second role for assessment is certification. Some systems base selection decisions on assessment which certifies completion to a required standard of a course or cycle of study (e.g. university entrance in England is based on GCE Advanced [A-]level results). Other systems base selection instead on assessments designed by or for the educational institution or occupation to which the candidate seeks entry (e.g. the Indonesian University Entrance Examination; the National Unified Entrance Examination in China). Meng Hong-wei (Chapter 9) describes assessment reforms in Shanghai in 1985 in which, for the first time in Chinese education history, educational achievement was assessed and certificated at the end of senior secondary school independently of the selective assessment for university entrance.

The certification role of assessment is often undermined by the dominance of its role in selection. Fouad Abou-Hatab and David Carroll (Chapter 11) describe a plan in Egypt to introduce a secondary school certificate which would emphasize certification and completion of secondary studies delinked from the university entrance examination. Intended to de-emphasize the negative aspects of the selective function of secondary education, the introduction of the certificate was resisted by public opinion.

A third role of assessment is the promotion of learning through systematic diagnosis and feedback of assessment data to teachers and students. This is the formative role of assessment and has characterized some types of routine teacher assessment for centuries. However, the use of assessments designed by outside agencies for use by teachers and other professionals began to accelerate in the 1930s as the intelligence and psychometric testing movements in Europe and the United States gained ground. Diagnostic tests were used most often in the diagnosis of learning problems among low achieving students but also for educational and vocational guidance and, especially in the United States, as an integral part of 'mastery' learning strategies.

More recently, some countries have developed mass assessment systems intended to provide information to help teachers teach better. Tony Somerset describes how in the 1970s Kenyan teachers were presented with analyses of performance on the Primary 6 selection examinations to inform their teaching of future cohorts of primary school students (Chapter 16). And Jahja Umar (Chapter 14) discusses how teacher assessment of student performance in Indonesia promotes learning.

These three roles—selection, certification and learning—have particular significance for the learning and life chances of individuals. In addition assessment is used increasingly as an instrument of system reform monitoring or system management. The monitoring and management of education systems involves a wide variety of assessment activity. Sometimes the monitoring or assessment of performance is undertaken to determine the causes of differences in academic achievement; sometimes to detect trends in achievement over time; sometimes to increase the financial accountability of the education system; and sometimes to evaluate the efficiency and/or effectiveness of educational innovations.

The growth of assessment for monitoring and management is especially apparent in industrialized countries. George Madaus and Anastasia Raczek (Chapter 6) discuss the phenomenal growth in assessment in the United States and the proliferation of the uses to which assessment results are put. The use of standardized tests as a policy tool for educational reform is singled out as a significant and recent development. Patricia Broadfoot and Caroline Gipps (Chapter 8) describes the National Curriculum Assessment System introduced in England and Wales in 1988.

Growth in monitoring is also apparent in those systems which receive financial and technical assistance from external sources. Marlaine Lockheed's account of World Bank support for educational testing (Chapter 2) describes how the bank did not make a single loan for educational assessment between 1963 and 1974. By the early 1990s around 40% of education projects included assessment components, the sharpest growth being apparent in projects which support the monitoring of progress towards national goals. The Quality of Educational Assessment System in Chile (known better by its Spanish acronym SIMCE) is an example of a monitoring project supported by a loan from the World Bank. Josefina Olivares (Chapter 7) describes how the system guides decision-making and tasks undertaken at different levels of the school system. Several international organizations are involved in promoting the use of assessment for monitoring. Vinayagum Chinapah (Chapter 3) describes the UNESCO/UNICEF joint monitoring project for primary education while Meng Hong-wei (Chapter 9) describes the recent involvement of China in this and other international projects.

Role conflicts

All the authors in this book address the question of conflict or tension between the various roles of assessment. Can assessment designed and used for one purpose fulfil another also?

The tension which has received most attention in the literature to date has been that between assessment for learning and assessment for selection, or, more precisely, that between learning in general and assessment for selection. In 1911 a Consultative Committee on Examinations in Secondary Schools in England and Wales summarized the 'good' and 'bad' effects of examinations on pupils. These included inciting the student 'to get his knowledge into reproducible form and to lessen the risk of vagueness', but also that 'by setting a premium on the power of merely reproducing other people's ideas and other people's methods of presentment' examinations divert energy from the creative process. The examination system in Imperial China also was not without its critics. In 1043 Fan Zhong Yan highlighted the stultifying effects of these examinations on imagination and their reinforcement of a derogatory attitude to studies with practical utility (Cleverley, 1985).

The tensions are contemporary. George Madaus and Anastasia Raczek (Chapter 6) review briefly the assertions about the positive and negative effects on the curriculum of standardized tests in the United States. Teaching to the test, test practice, and the neglect of educationally important, non-tested skills are among the negative effects cited.

Possible conflicts between assessment for learning and assessment for monitoring have also been identified. Elsewhere Nuttall (1993) has suggested that the dual aims of the National Curriculum Assessment in England and Wales—to provide diagnostic information on individuals for teachers and judgemental information on schools for the general public— are contradictory. Teachers will be motivated to secure high student scores not by a desire to understand better a student's learning difficulties, but by a desire to appear to be a successful teacher. In other words the national curriculum assessments are 'high stakes' for teachers. On the other hand, a 'low stakes' monitoring test runs the risk of bringing out the best or the most representative in neither student nor teacher. Similarly, Goldstein (1991) has argued that a system designed to diagnose individual learning weaknesses must encourage 'openness' and 'honesty' so that weaknesses can be appraised and ameliorated. Systems designed to judge schools or teachers will discourage the identification of weakness.

Some have suggested that using assessments designed for one purpose for another purpose is always inappropriate; that it is effectively impossible to pursue two objectives at once. Marlaine Lockheed (Chapter 2) argues that the purpose of an assessment determines its design. Such discussions often refer to an assessment's 'fitness for purpose'. Fitness may involve features as different as item difficulty, the range of material assessed, the format of the assessment, the standardization of context, and the assessment's reference to the performance of others or to performance criteria. For example, selective assessment requires that items discriminate sharply between individuals, often narrowing the range of competencies tested. Assessment for monitoring the performance of a national system, by contrast, is designed to assess a wide range of competencies and does not need to discriminate finely between individuals.

Others suggest that there can be a degree of resolution between the various roles of assessment. Caroline Gipps (Chapter 15) acknowledges fitness for purpose but argues that any assessment for selection, monitoring or accountability which involves a complete population, rather than a sample, of students can and indeed must use a model of assessment which enhances and supports learning. School-based teacher assessment must be trusted; teachers must be trained in observation, diagnostic questioning and formative assessment, curriculum definition, performance exemplars and various processes of moderation; and external assessment should be limited to one or two ages only.

The reader should bear in mind that Gipps' confidence that roles can be resolved is rooted in an English educational context in which there is a long and largely unbroken tradition of trust in teacher judgements and school-based examining. The absence of such a tradition in some contexts is demonstrated in Sunderi Kariyewasam's account of the deep mistrust of teachers' judgements in Sri Lanka (Chapter 13). When the Sri Lankan Examinations Department attempted to incorporate teacher assessment of project work into the results of important public selection examinations in 1988 resistance was intense and the overthrow of the scheme rapid.

This brings us back to the issue of context and an acknowledgement that the terms of the assessment debate are rarely universal for policy makers, parents, teachers or students at any given point in time. Historic, economic and political characteristics will shape the pattern of dominance of assessment's roles.

The organization of the book

The book is divided into four main parts. In the first, trends in and pressures on assessment policy and practice are set in global, national and historical context. Three of the chapters adopt a comparative approach to explore the value, possibilities and limitations of establishing universal propositions about the antecedents and consequences of assessment practice or policy. Two of the chapters describe the efforts currently being made by international agencies to influence the introduction of assessment as a tool for educational policy reform by national bodies.

In the second and third parts, assessment systems are examined in national contexts through country case studies. Clearly, it is impossible in a single volume to survey every country grappling with assessment issues. Most of the world would be represented. The country case studies are selected either because they represent different approaches to large-scale assessment reform (Part two) or because they demonstrate the different ways in which educational, economic political and cultural environments influence the dominant and emerging roles of assessment (Part three). Part two draws examples of large-scale assessment reforms from the United States, Chile, England and Wales, and China; while Part three explores the tension between the selection, learning and monitoring roles of assessment in South Africa, Egypt, Korea, Sri Lanka and Indonesia.

In Part four the possibilities and limits of tension resolution are explored in three theoretically oriented chapters which tackle, respectively, the need to subordinate the purposes of assessment for selection, monitoring and accountability to a model of assessment for learning; the relationship between examinations and educational quality; and the value of the concepts of individual choice, incentives and control in the understanding and resolution of assessment dilemmas. Finally, the afterword summarizes the lessons of the book and the evidence for and against global convergence in assessment practice.

References

Cleverley, J. (1985). The Schooling of China. Sydney: George Allen and Unwin.

Goldstein, H. (1991). Assessment in Schools: an alternative framework. Education and Training Paper No. 5. London: Institute for Public Policy Research.

Nuttall, D. (1993). Monitoring National Standards. Research Working Paper No. 6. London: International Centre for Research on Assessment, Institute of Education.

World Declaration on Education for All (1990). Framework for Action to Meet Basic Learning Needs. Jomtien, Thailand.

A.L.
A.W.

Part one
GLOBAL TRENDS AND PRESSURES

Part one provides a broad overview of the state of educational assessment worldwide, of the trends that can be identified in varying degrees across the globe, and of the underlying pressures—often economic—which manifest themselves everywhere. In Chapter 1 Angela Little uses the cases of Sri Lanka and England to explain the importance of both the general and the particular. Countries' assessment systems operate within an increasingly interconnected world, and face pressures which are shared everywhere, notably those for selection, monitoring, and the improvement of learning. At the same time, each country's history is unique. In order to understand how assessment operates in a country and how, therefore, it might be improved, it is not enough to look at general trends. As Little emphasizes, 'national contexts', the particular, the specific and the cultural are also vitally important.

The authors of Chapters 2 and 3 work for two of the international agencies which are now so important in the development of assessment policy. Marlaine Lockheed explains the philosophy behind the World Bank's increasing volume of lending for assessment systems; while Vinayagum Chinapah explains how UNESCO is trying to make concrete the objectives identified at Jomtien.

In Chapter 4, Harvey Goldstein provides a discussion and analysis of international comparisons of achievement. As these have become increasingly popular, with more and more countries participating, and also better known to the national and international press, so their potential impact on educational policy has increased. Goldstein's chapter discusses our experience to date with this avowedly international form of assessment, and evaluates its potential, and its limitations.

Finally, Chapter 5, Harold Noah examines whether lessons can be learned from comparative research on assessment policies and practices, and whether comparative research is an essential component in the making of assessment policy. While the primary lesson is that there is no single best system of assessment, and that many national systems have, in any case, been resistant to foreign imports, a number of common themes and important distinctions are identified.

CHAPTER ONE

Contexts and histories: the shaping of assessment practice

Angela Little ·

Nageswary is 11 years of age and attends school in Sri Lanka. Although she will not sit for her first major selection examination until she is 16 years of age she has been attending private tuition since she was seven – two hours each evening, including weekends, for lessons in all her school subjects. Her parents pay for these lessons and for those of her brother. Approximately 10% of the household monthly income is spent on these classes. Her teachers frequently set tests and she has to study hard for them. She has recently sat the primary Grade 5 scholarship exam but failed to achieve the marks which would secure her a modest financial bursary and access to a 'good school'. However, this failure will not prevent her from continuing her schooling. In Sri Lanka the 'open access span', or the number of grades to which students may proceed before sitting a public selection examination, is 11 years of education, or more if the student repeats a grade. Nageswary has little time for play with her friends. She is either studying or helping around the house, fetching water, looking after her young brother, washing dishes or fetching firewood. The General Certificate of Education (GCE) Ordinary (O-)level examination which she will take at the age 16 is extremely important for her. Without six good passes, preferably with credits and distinctions, her chances of a 'decent job' with a regular wage or salary are slim. The unemployment rate among those with GCE O-level qualifications is high, and for those with GCE Advanced (A-)level qualifications even higher.

Ruth is also 11 and attends school in England. She will also sit for her first major selection examination when she is 16. Although her teacher sets few tests for the whole class, Ruth's work is frequently assessed on an individual basis. Ruth's class has also just participated in a series of assessment tasks which the government uses to monitor standards in education across the country, and which are set for all 11-year-olds. Her teachers are concerned about these for they will reflect on the status of the school as a whole. Ruth has about one hour's homework each evening and a little more at weekends. She does not attend any extra classes nor do her friends. After school they play at each other's houses and on some evenings go to swimming and dancing lessons together. Her parents spend approximately 1% of their monthly income on these extra activities. The General Certificate of Secondary Education (GCSE) examination which Ruth will take at age 16 will be very important for her but her 'life-chances' will not be determined to quite the same degree by educational qualifications as those of Nageswary. The stakes are certainly high for Ruth, but much higher for Nageswary.

The role of educational assessment in the lives of young people in different countries bears a surface similarity. Beneath the surface are major differences in assessment's significance: for life-chances, for the types of learning which are encouraged, for the value placed on it by teachers and parents and for the personal investment which parents make in improving children's performance. All of these, along with students' attitudes to assessment, are embedded in national and local systems of education, assessment and employment.

National and local systems are themselves embedded in labour markets, communication systems, qualification systems and commercial assessment activities which are increasingly international in orientation and control. Some of these international influences are bilateral and involve a single other country; others are multilateral or multinational and involve interests which cross-cut several nations. International influences interact with national and local influences to varying degrees. The education and assessment systems of some countries, notably those in the industrialized north, are relatively immune to external bilateral and multilateral pressures. The systems of the south, dependent on external financial cooperation and technical assistance, are more vulnerable to external pressures. This too creates differences in the role and significance of assessment: differences of which this chapter provides an overview.

This chapter explores the value of analysing educational assessment in international, comparative and historical context in three ways. First, it describes the advances in knowledge from studies which move beyond single contexts. Second, it examines how different contexts accommodate superficially similar assessment systems, through a review of assessment in two countries whose histories are themselves interwoven—Sri Lanka and England. Finally, it explores the value and limitations of comparative propositions about assessment. While this chapter explores the value and the limitations of international, comparative and historical studies of assessment, subsequent chapters will explore emerging international trends in assessment or contemporary assessment practice in single country contexts.

The value of international, comparative and historical studies

Comparative and international studies in assessment offer a critique of studies bounded by time and space. Wallerstein's (1993) comment on comparative sociology to the effect that 'it is not a field but a critique of whatever seems narrow' is apposite. International and comparative studies of assessment help us acknowledge the limits of studies conducted in single contexts and invite at least four advances in knowledge—locating the bounds of time and space, extending variance, perceiving invariance and appreciating system interdependence.

Locating the bounds of time and space

One of the best examples of how knowledge of assessment and achievement generated in one context is transferred to other, often radically different contexts, is to be found in the literature on the effects of school and home on the assessment of academic achievement.

In the 1960s and 1970s, studies from the United States emphasized the relatively greater importance of home over school effects in the explanation of differences in academic achievement. So powerful were the arguments and compelling the evidence that the proposition was on its way to becoming a universal orthodoxy. Some years later, the size of school effects on achievement in London in the late 1970s/early 1980s was re-evaluated using sophisticated multilevel modelling techniques (Rutter *et al.*, 1979; Mortimore *et al.*, 1988). The effect of school variables was found to be much greater than suggested previously by the US studies. This difference may have arisen for one or more of several reasons—the variation in both home and school variables in the London school sample compared with the US sample, the characteristics of English or US society in general, the use of different statistical techniques, or changes over time, which may or may not have been common to both societies, but disguised by the different times at which the studies were undertaken. Which of these reasons best explains the differences found has not been evaluated but the evidence highlights the need for caution in the transference of results from one context to another.

In fact, studies of academic achievement in developing countries in the 1970s had already suggested that school effects may be very strong in societies which vary in the degree to which their schools provide differential resources and support for learning. The relative effect on achievement of school and home factors will itself depend to some extent on the amount of variation among schools and among homes in the population selected for study. Schiefelbein and Simmons (1981) and Heyneman and Loxley (1983) provided early reviews of the effects of home and school on academic achievement in developing countries and, contrary to the studies from the United States, showed strong effects of school factors on academic achievement. Although a later study from Zimbabwe, using multilevel techniques, questioned the strength of the school effects (Riddell, 1989), the momentum built up by these early studies in developing countries had generated a view which ran counter to that circulating in industrialized countries. While the transference of research-findings in countries in the north to those in the developing south has been slowed by the results of these studies, they in turn began to be transferred for policy purposes across a range of potentially different contexts within the south.

Extending variance

One of the advantages of examining assessment in a range of countries is that the variance of economic, political and social factors which affect assessment is extended. Education systems vary enormously across time and space but the systems of industrialized countries bear many similarities, not the least of which are compulsory education cycles for students to their mid–late teenage years, class sizes of rarely more than 50 students, and welfare systems which support the young unemployed. Each of these will affect the motivations and learning practices of students as well as their teachers and parents. When studies of learning assessment are undertaken among populations of students, teachers and parents who act within these system characteristics, we do not know how far we can generalize their results to systems whose characteristics fall beyond this range of variance. Class size is the classic example. Many studies from industrialized countries suggest that variations in class size between 25 and 40 students make little difference to levels of student achievement, and lead policy analysts to the spurious policy conclusion that 'class size is unimportant for achievement'. In some developing countries class sizes can reach 200, extending the variance of class size considerably. However, few studies have been undertaken on the effects of class size on achievement in such systems. Had they been, then it is likely that our understanding of the effects of class size on achievement worldwide would be extended considerably.

Perceiving the invariant

Comparative studies can also throw into sharp relief the invariant and taken-for-granted features of an education system. Apprehension of causality in the social sciences tends to be based on associations between variables *within* national boundaries. Factors which do not vary tend to be excluded from consideration, in part because it is difficult to demonstrate links between factors which are common and factors which vary. However this does not mean that only factors which vary are important.

For example, when all children within one education system are taught and assessed at home and school in the *same* language, the mother tongue, language of learning and language of assessment may not be included as variables in models of school achievement or school effectiveness. When children in a second education system learn at home and school in different combinations of language, one cannot fail to be struck by the language and learning difficulties which children in the second system face. But because the difficulty may be common to all children in the second system it may not emerge as a key variable in the explanation of differential achievement within that system.

Only when variables which distinguish between national systems are included in comparative analyses may one begin to understand the taken-for-granted and invariant features of both systems and their effects on students' achievement. Common rather than differential achievement within a country is often appreciated only when seen in relation to differences between countries. In other words, comparisons with a second system can throw into sharp relief the invariant features of the first and can encourage an understanding of the factors which lead to the common, rather than differential characteristics of student learning.

System interdependence

A fourth advance in knowledge generated by an international approach is an understanding of the impact of forces external to a country on those which are internal. Some studies examine the impact on national systems of assessment of external economic and educational forces for change (e.g. Lewin and Little, 1984; Sancheti, 1984). These were influenced by 'dependency theory', a Latin American inspired theory of economic underdevelopment which was extended to include ideas of neo-colonial cultural and educational dependency. Most significantly, for international and comparative education, the ideas of dependency theory encouraged a shift in the unit of comparative analysis, away from the nation state and towards *relations between or interdependence of* nation states.

Twentieth-century social science accounts of educational and social change in industrialized countries have tended, until very recently, to focus on explanations which draw on national and intranational forces. In the late 1960s and early 1970s however, significant conceptual advances in development economics flowed over to analyses of education in developing countries. The dependency perspective focused on underdevelopment rather than development, viewing it as a necessary outcome of systematic exploitation and manipulation of peripheral economies by central economies (Frank, 1967; Cardoso, 1972; Dos Santos, 1973). It was argued that poor countries, conditioned by their economic relationships with rich economies, occupy a subordinate and dependent role which inhibits development by expropriating investible surplus. Indigenous élites, firmly wedded to the international capitalist system and rewarded handsomely by it, have no interest in giving up these rewards, and support education and qualification systems which maintain access to them. Dependency theory accords overriding importance to the historical conditions which provide a context for development and to the international system of 'global exploitation' managed by developed capitalist countries. The 'dependency' perspective encouraged economists, political economists and sociologists to abandon the national economy, nation state and national society as their central unit of analysis and to focus instead on the nature of relations between economies, states and societies. Dependency is conceived as a cultural phenomenon also. The structure of dependent economic relations was asserted to create a 'cultural alienation' in which values, norms, technology, concepts and art forms were inspired externally rather than internally (Carnoy, 1974). Such analyses in the 1970s would lead, by the 1980s, to accounts of world systems of knowledge production (Galtung, 1980) and cultural globalization (Featherstone, 1990), which are not necessarily dependent on parallel processes of economic globalization (Waters, 1995).

Contemporary accounts of assessment and educational change in all countries, but especially in former colonies, began to take account of economic and cultural forces at work beyond the nation state, enriching understanding. For example, agencies which are international rather than national in staffing composition and funding represent an increasingly important influence on national systems of education and assessment. Not only is the volume of international financial lending for assessment projects increasing, but it is clear that the pattern of spending and types of projects supported are changing and the conditions attached to the monitoring of that spending more focused. International or global labour markets, mediated by perceptions of how access to these may be gained are a second important influence. As we shall see both types of influence have been important in shaping the course of assessment practice in Sri Lanka.

National contexts and international histories

A comparison of assessment in Sri Lanka and England illustrates this chapter's main thesis: namely that the roles of assessment in education should be understood in relation to national and local educational and social context, to the history of educational assessment and to the dependence of some systems on external influence. This comparison also provides a basis for understanding some of the differences between Nageswary and Ruth's experiences, described at the beginning of this chapter, and the more general implications of assessment practice for children's lives.

Foreign influences on education and assessment: Sri Lanka (formerly Ceylon)

Formal education in Sri Lanka has a long history. Buddhist, Hindu and Muslim schools predated the European colonial influences on education by many centuries. Each educational tradition had its own assessment practices. Promotion within the Buddhist temple schools, for example, was dependent on performance at three distinct levels. Successful performance in tests devised by superiors and based mainly on the knowledge and understanding of religious texts resulted not only in promotion through the ranks but also in increased material rewards (Rahula, 1956).

European colonization of Ceylon began in 1505 with the capture of the maritime coastal areas by the Portuguese. The colonial state religion was Roman Catholicism. By the end of the Portuguese period in the mid-seventeenth century an extensive system of parish schools attached to Catholic churches had been established in the coastal areas. There were secondary schools and colleges run by Jesuits and Franciscans and a Franciscan teacher training college (Ruberu, 1962). Education was under the control of the different orders of the Catholic Church and was oriented to religious conversion. Sinhala or Tamil were the instructional languages and Portuguese and Latin were taught as subjects in the colleges. Indigenous Buddhist, Hindu and Muslim schools were officially discouraged.

The Dutch displaced the Portuguese in 1658 and colonized the maritime provinces of the island until 1796 through an administration run by the commercial Dutch East India Company. The Dutch continued the tradition established by the Portuguese of using centres of learning for religious conversion, but promoted vigorously the Protestant religion of the Dutch Reformed Church and discouraged the Catholic faith of the Portuguese. They extended the system of parish schools established by the Portuguese in the areas of Colombo, Galle and Jaffna. All parish schoolmasters were required to profess the faith, as were others who sought office in the colonial administration. Sinhala or Tamil were used as the media of instruction. No attempt appears to have been made to teach the Dutch language in the parish schools, though it appears to have been taught to prospective teachers and preachers at the Colombo Seminary alongside Latin, Greek and Hebrew.

Sri Lanka's relations with England were established when the British replaced the Dutch as the colonial power in 1794. Educational and assessment traditions based on Buddhism, Hinduism and Islam had been undermined by the Portuguese and the Dutch in the eighteenth century and replaced by modified forms of the educational practice then current in their respective home European cities. The British continued the process of exporting metropolitan educational models, resulting in an education system which reflected a number of different European traditions in education, all of them differentiated sharply from the traditional Buddhist, Hindu and Islamic forms.

Growth of the colonial English education system was slow in the first half of the nineteenth century but quickened in the second. This was a time when job allocation based on nomination and patronage was being eroded in England. Examinations for civil service posts abroad (especially in India but also in Ceylon) were introduced and English universities were establishing the examination boards which would in later years control school examinations. The first public examination (8th Grade) was held in Ceylon in 1862. Although it was 'only a copy of what was going on at the time in England' it was set and marked by the Department of Public Instruction in Ceylon (GOSL, 1972). The examination survived for 18 years before it was replaced by the Cambridge Examination in 1880 and the London Matriculation in 1882. In 1883 a Ceylonese candidate was placed twenty-seventh in competition with candidates from England, India, Ceylon and other colonies, and first in the India and Ceylon list. The award of the Gilchrist scholarship to this candidate for further studies at London or Edinburgh created for Ceylonese the precedent of access to further education across national boundaries. Educational assessment held open the possibility of international mobility for the lucky few.

The need of the colonial administration for Ceylonese trained for white collar jobs provided the rationale for education in colonial Ceylon. Success in life came to be associated with success in school examinations. Those who controlled the colonial system (British colonial civil servants) had no major personal stake in the Ceylonese schools and examinations, which had little effect on the education and occupational success of their own children. Examinations were used to legitimize, via achievement criteria, the creation of a colonial élite which was to replace the traditional, ascriptively defined rural élite. This did not mean the wholesale replacement of one social group by another. In the low-country areas those Ceylonese who took early advantage of English education and examinations and gained government jobs were from already élite low-country families. In the up-country areas, by contrast, the Kandyan élite resisted British colonization and education. Their lack of educational success was used by the British to undermine their access to positions of power under the colonial regime. Over time the colonial indigenous élite became the post-independence national élite. Those who controlled the education system now held a family stake in the competition for white collar *and* élite government positions. Since their own positions and those of their parents and grandparents had been gained through success in examinations they were committed to maintaining an education system oriented to assessment for selection.

The early school examinations in Sri Lanka were stratified by language medium. English medium schools and examinations enjoyed a much higher status than 'vernacular' (Sinhala and Tamil) schools and examinations. Success in the English medium examinations provided access to the best government jobs open to Ceylonese. Gradually, success in the vernacular examinations provided access to lower-level government jobs, especially in teaching. Success in life was closely related to success in examinations, especially success in English medium examinations.

From a dual to a unitary system of assessment: Sri Lanka

The 1945 education act in Ceylon was passed just three years before independence. Education from primary level to university in the state-supported system would be free henceforth. From a dual system of education based on language and class Ceylon moved towards a unitary system of education and a unitary system of assessment. Only a handful of the élite English-medium schools renounced state subsidies and remained outside the state system.

The language of instruction as well as the language of government administration remained a politically contested issue. Whereas English had become the language of government and of education among the colonial élite, the rest of the population studied in either Sinhala (the language of the majority) or Tamil (the language of the minority). The success of the Sinhala nationalist Sri Lanka Freedom Party in 1956 led to Sinhala being proclaimed the official language of government. The official medium of instruction in élite schools had already begun to change from English to either Sinhala or Tamil. The change was introduced into the 1st Grade of primary schools in 1945, in secondary in 1953 and in the universities in 1960. The language of examinations followed suit. The earlier dualism of the education and examination system stratified by language (English versus one of the vernacular languages) was replaced by a single state-supported system *differentiated by language* (Sinhala versus Tamil).

growth and increased competition for jobs

'50s success in school examinations had guaranteed
..o government positions. But the development over time
.. free education and examination system (albeit differentiated
,uage medium) was not matched by a substantial growth in the
..onomy and growth in the types of jobs which people had come to
associate with examination success. Jobs became scarce while certificate
holders proliferated. Mass participation in the education and examination
system was not matched by mass participation in the economy. Nor did
growth of the university system match growth of the secondary school
system. The transition from a dual to unitary system of education and
assessment occurred in an economy which was not expanding rapidly
enough to absorb the numbers of qualified students who would subsequently emerge. The pressures to succeed in the competition for scarce
resources in both education and employment increased.

An absence of foreign definitions: the English context

Even though educational and assessment practice in Ceylon before independence was modelled on English practices, the contrasts between the
underlying systems of education and society were striking. The education
and selection system in England experienced slow and continuous change
during the late nineteenth and early twentieth centuries. Public examinations for secondary school students were first introduced in England in
1857 through a collaboration between Exeter University and surrounding
schools. Oxford and Cambridge Universities followed in 1858 and London
a few years later (Mortimore and Mortimore, 1984). These initiatives
grew out of a number of early-nineteenth-century developments which
included the replacement of religious admission tests set by Oxford and
Cambridge Universities by tests assessing a broader base of knowledge, the
introduction by the established professions of qualifying examinations, and
the introduction of the competitive examinations for entry to the colonial
and later the home civil service.

To date, England has escaped attempts by foreigners to define educational purpose and the criteria of educational success, though she has borrowed ideas from Europe and the United States from time to time.
Definitions of educational purpose are contested but the debate involves
interest groups internal to the country not external. The diverse and highly
decentralized structure of English education in the nineteenth century gave
way only slowly to an integrated system which has retained a dualism based
on social class and the ability to pay. Private schools are exempted from

National Curriculum requirements and the Standard Assessment Tasks used for system monitoring, though those admitted to universities from the private sector follow GCE A-level examination courses because the universities use them for admission purposes. Indeed in some respects the control of contemporary assessment policy in England resembles that of nineteenth-century Ceylon. Those in control of education and assessment policy in England educate their own children in schools which are exempted from state control. This was also the case in colonial Ceylon where those who prescribed colonial state policy placed their own children in schools which lay outside colonial state control. This is not generally the case in contemporary Sri Lanka where educational policy makers educate their own children in state schools, albeit 'good' state schools. The establishment in England in the 1960s of a two-tier system of public examinations at 15+ (the GCE and the CSE) differentiated broad ability groups in a way which also tended to reflect broad social class groupings. This differentiation has never been used in Sri Lanka where all students who reach Grade 11 sit the same public examination.

There are other contrasts. The decentralized definition of curriculum and the plethora of examination bodies in England stand in contrast to the early unified curriculum and assessment guidelines set out by the Department of Public Instruction in Ceylon, established in 1869. In England the language of instruction and examinations and the language of administration have never been central political issues. Nor has the issue of educated unemployment destabilized the British government to the extent that it has in Sri Lanka over the past two decades (though they have been in Wales and the periphery of the British Isles). Access to university in Sri Lanka is far more difficult than in England. The sense of frustration borne of blocked access to future opportunities, mediated in part by selective assessment in Sri Lanka, has resulted in three episodes of youth insurrection over the past 25 years, with an estimated loss of over 100,000 lives.

The different educational and social contexts of Ceylon and England have accommodated superficially similar assessment systems in different ways. Ceylon introduced school examinations in the mid-nineteenth century at the same time as England. In Ceylon they were externally imposed in a foreign language and were used to ration access to a limited number of colonial government jobs. They were also used to legitimize the creation of a colonial élite which was to become the national élite in the post-independence era.

In England examinations arose out of domestic concerns about the religious control of educational institutions, the need of professional bodies to restrict access to membership, the rapidly growing colonial and home civil service, and growing opposition among the Victorian middle classes to patronage and aristocratic privilege gained through 'jobbery'. Examinations were conducted in the native language and were used increasingly into the twentieth century to ration allocation to further education and jobs. Although assessment for selection is still extremely important for teachers and students, the gradual expansion of the senior high school (the sixth form), and the diversification of assessment at this level, are reducing some of the pressures exerted by the examinations at 16+. The newer assessments faced by students lower down the system are national assessment tasks designed to monitor the performance of the educational system as a whole rather than to select students for further education or jobs.

The older aristocratic forms of patronage and reward have continued in England and have not been wholly replaced. In Ceylon the contemporary élite was to inherit the attitudes towards merit forged by the English Victorian middle classes. Max Weber's thesis on the bureaucratic character of 'late developing societies' with their rationalization of rules, means and ends (of which the certificate is a manifestation) applies well to Ceylon. The certificate replaces family connections as a prerequisite for advancement; it confers social prestige; it supports claims for marriage into notable families, claims for respectable remuneration and assures advancement. It supports 'above all, claims to monopolize socially and economically advantageous positions' (Weber, 1972: 226).

This is not to say that family connections are unimportant in Sri Lanka where there is now a considerable degree of intergenerational transmission of status via education. Better-off families can guarantee better and more education for their children. But poor families and poor children in Sri Lanka are more likely to believe that they too can achieve high status through educational and examination success than their peers in England.

The asymmetry of the assessment relationship between Sri Lanka and England

Sri Lanka's history of educational assessment cannot be understood in isolation from an understanding of assessment history in England. The reverse is not true. The history of assessment in England can be understood in isolation from that in Ceylon, except in respect to the growth of job opportunities for young English men in Ceylon in the mid-nineteenth century and the need to find a rational and legitimate means of selecting such persons. The development of educational school assessment in Sri Lanka was generated by outsiders in response to their needs; and in England by insiders in response to domestic needs. In Sri Lanka achievement criteria replaced ascriptive criteria for occupational allocation more completely than they did in England, and at a much earlier relative stage of development of the education system as we know it today. Changes in assessment practice in Sri Lanka continue to be influenced by changes in practice in England, though the antecedents and consequences of change are often very different in the two contexts: the reverse is not true. There is also another abiding and important difference in the context of educational assessment, namely the size of the imbalance between qualified job seekers and available jobs. In Sri Lanka this greatly reinforces the value attached to assessment for selection.

This final difference and its counterpart elsewhere, has led some writers to move beyond historical and cultural analysis and to attempt a more systematic comparative analysis of educational assessment. Contextualized studies are a necessary first step towards an international and comparative analysis of assessment. Whether the next step, involving the generation and confirmation of cross-national propositions which have predictive power, can be undertaken, is, as we shall see in the next part, open to debate.

Towards comparative propositions about assessment

One of the more controversial and systematic comparative analyses of assessment and its implications for education is that conducted by Dore (1976) who focused on the incompatibility between assessment for selection and the achievement of the educational goals of relevance and independence in learning.

> Not all schooling is education. Much of it is mere qualification earning. And more and more of it becomes so. Everywhere, in Britain as in India, in Russia as in Venezuela, school is more often qualification earning schooling than it was in 1920 or even in 1950; ritualistic, tedious, suffused with anxiety and boredom, destructive of curiosity and imagination; in short anti-educational. (Dore, 1976: ix)

The 'anti-educational' assertion about selective assessment provided a cornerstone for an elaborate thesis, the 'diploma disease', about the relationship between labour markets, educational certificates, the quality of the process of learning and teaching and the potential of schools to develop 'relevant' skills. The thesis was heavily influenced by the educational and economic experience of 'late developing' countries which began their drives for modernization in the late nineteenth or twentieth centuries.

> The later development starts (i.e. the later the point in world history that a country starts on a modernization drive) the more widely education certificates are used for occupational selection; the faster the rate of qualification inflation and the more examination oriented schooling becomes at the expense of genuine education. (Dore, 1976: 72)

The thesis stands in contrast to functional sociological perspectives on assessment which had viewed the school as the producer of skills required by society, assessment as the efficient allocator of persons to different occupational roles, and curriculum and assessment reforms as responses and adaptations to change in the economy and occupational structure. Dore's thesis is that the 'functional' roles of the school and assessment are not in fact played out in many developing societies because of an underlying imbalance between the number of occupational roles and the numbers of qualified competitors for them, and the consequent constraint placed by assessment on the production of relevant skills.

Discrete elements of this thesis were not new. For some years economists had been noting the trends in industrialized and in industrializing countries towards educational inflation or 'certificitis' (e.g. Bowman, 1970; Milner, 1972). Educationalists had highlighted the positive and negative 'backwash' of examinations on the teaching and learning process for much longer (e.g. the Morgan committee in Ceylon in 1867; the consultative committee on examinations in secondary schools in England and Wales in 1911).

The novelty of the thesis lies in the creative synthesis of these issues within a comparative sociohistorical framework embracing concepts of bureaucracy, economic and social dualism, and societal 'late development'. The thesis weaves its way from labour market structures, and income, status and prestige differentials to recruitment practices for salaried employment, excess demand for higher education and an over supply of educated graduates. In turn these lead to qualification escalation, increased public demand for more and higher levels of education, selection-oriented assessment, and selection-oriented teaching and learning in schools. This last, Dore argues, controversially, leads to the underdevelopment of abilities, the 'deformation' of character and the thwarting of attempts to provide relevant education for all, and, in consequence, leads to the slow development of the economy.

Two propositions fundamental to the general argument are (i) that there is a conflict between assessment for selection and curriculum objectives, and (ii) that there is a conflict between learning motivated by selection and learning motivated by interest in the task or subject matter being learned. For the general thesis to have cross-national validity these propositions require confirmation within each country studied. The thesis also predicts that the symptoms of the diploma disease intensify over time. In the following section the validity of the two assumptions and the prediction are examined.

Assessment for selection and curriculum objectives

Systematic comparative research which would confirm worldwide the conflict between assessment for selection and curriculum objectives has never been conducted. One of the more systematic attempts to examine the tension between assessment for selection and curriculum objectives was made by Lewin (1984) in his examination of the curriculum development process in integrated science in the mid-1970s in Malaysia. The Malaysian Integrated Science (MIS) reform, designed for students in Grades 7–9, attempted to decrease the use of didactic teaching methods and increase the use of more heuristic approaches to learning. Science was to be approached through an exploration of the world of experience rather than as a body of factual information to be memorized. Active learning was to be encouraged and attitudinal outcomes given considerable emphasis (e.g. interest and enjoyment in scientific activities, awareness of the social and economic implications of scientific activity). The end-of-course assessment was intended to reflect these curricular emphases and in particular to examine an understanding of practical techniques, the design of experiments and solution of problems through experimentation. The end-of-course assessment was part of a selective examination in which approximately 60% could expect to pass and proceed to Grade 10.

Analysis of the 1972–5 examination paper suggested that in each year's paper over half the items should be classified as knowledge (i.e. recall or recognition) level questions. There was no provision for any practical examinations. Few questions tested the application of knowledge and none assessed affective outcomes. Responses from 105 school teachers involved in the programme suggested widespread dissatisfaction with the effects of the examination. Teachers commented that the questions used , which were multiple choice in format, inhibited expressive ability. They claimed that pupils were only interested in studying for the examination, that teachers concentrated on facts and discarded discovery methods and that the examination questions did not reflect adequately the curriculum objectives.

Despite teachers' largely unfavourable perception of multiple-choice tests they appeared to favour this technique in their own methods of internal school assessment. Seventy percent of teachers reported that they used a multiple-choice format for more than one half of all assessments given. Sixty-six percent reported that they had never used practical assessment of any kind (Lewin, 1984: 135–139).

Students' reactions to MIS and the end-of-course examination suggested that their attitudes to learning science were influenced by the exam. Students perceived that multiple-choice questions were easier than essay questions involving expression and believed that such questions were dependent primarily on recall. Classroom observation of 40 class periods in 15 schools suggested that less than 16% of class time was spent on students undertaking experiments. Most time involved the teacher addressing the whole class. A detailed analysis of these interactions showed that the majority of time (56%) was devoted to the introduction or recall of facts or principles.

In summary, there appeared to be a degree of conflict between the learning aims promoted by the curriculum and the effects of the selective examination. A Malaysian teacher remarked pithily:

> The purpose of the integrated science course is to develop the ability to observe and reason; the purpose of the school is to get as many examination passes as possible.

Learning motivation

The second proposition embodied in Dore's thesis asserts a tension between learning motivated by selection and learning motivated by task interest. This proposition is not new in itself and has frequently appeared in the literature arising from a US or English context.

Hextall and Sarup (1977) assert that assessment systems encourage a learning ideology of individualism, the reproduction rather than production of knowledge and the acceptance of hierarchy. Such learning transfers from school learning to out-of-school learning and 'reproduces' the structure of wider society. They suggest that students exchange the products of their labours for housepoints, grades and certificates and in so doing become alienated from the labour of learning, surrendering the intrinsic value of productive learning activity, and substituting an 'exchange-value' under the control of expert examiners. Learning becomes a commodity to be exchanged in the market of record cards, mark sheets and grades.

This sociological interpretation of the link between assessment for selection and learning finds echoes in experimental psychological work in the United States. Working with primary-school children Deci (1975) examines the consequences for motivation and learning of linking extrinsic rewards (e.g. grades) to learning which is driven initially by the desire to 'master' a task. He suggests that the association of extrinsic rewards undermines rather than enhances subsequent levels of 'mastery motivation'. Maehr's (1976) extension of some of this work introduced the term 'continuing motivation' and suggested that many analyses of assessment, motivation and learning ignore the long-term effects of one on the other.

Not all perceive a necessary conflict between learning for 'exchange' and learning for 'use'. Many years ago Allport (1937) argued that instrumental motivation of the kind represented by assessment grades could help develop intrinsic or internal motivation. This potential was termed 'functional autonomy' and lay behind some of the thinking advanced by

McNamara (1980) in Papua New Guinea in the early 1980s in relation to curriculum and assessment reform. He explained functional autonomy thus:

> We can use carrots to induce a donkey, which has never eaten sorghum, to eat sorghum and to learn to enjoy it, so that at some later stage when carrots are no longer available he will happily amble in the direction of sorghum which he would otherwise have ignored. In such cases the learning survived the extrinsic motivation of the carrots, so that the sorghum itself becomes intrinsically motivating. (McNamara, 1980)

Students' perceptions of learning motivation

If the links between assessment for selection and learning were similar in all countries then one might expect to find consistent relationships between students' perceptions of alternative sources of motivation. Students' personal views of motivations for learning have been explored in a cross-national research programme by the Students Learning Orientations Group (SLOG, 1987). Working with secondary school students in six countries the group distinguished three types of motivation—learning motivated by assessment, learning motivated by subject interest and learning motivated by 'significant others' (e.g. teachers, peers, parents). The relationship between these three varied between the six countries. For the Japanese and Malaysian students there was a very low but positive relationship between expressed assessment orientation and interest orientation. In England, India, Nigeria and Sri Lanka, by comparison, the relationship was strongly positive (SLOG, 1987: 37, Table 3).

An extension of this work in Japan confirmed this weak relationship between being motivated by external assessment and being motivated by interest (expressed in the Japanese study as 'instrumental' and 'intrinsic' motivation, respectively). The study explored links between motivation and Japanese notions of alienation (feelings of loneliness, emptiness, oppression and self contempt). The study empathy (feelings towards others) (Azuma et al., 1987). Instrumental motivation was positively related to feelings of oppression and restriction and a lack of sensitivity to others. Intrinsic motivation on the other hand was negatively associated with feelings of loneliness, emptiness, oppression, restriction, self contempt. It was positively associated with a sensitivity to others.

The SLOG research and the extension by Azuma et al. suggested that the tensions between sources of motivation take different forms in different countries for students at similar stages of educational selection. The

Japanese and Malaysian data approximated most closely the predicted lack of positive relationship between learning motivated by selective assessment and learning motivated by task interest. Data from England, Sri Lanka, Nigeria and India confirmed the prediction only weakly. However, the English students' pattern of perceptions was differentiated from all other countries with respect to learning motivated by the expectations of others. While the relationship between learning motivated by assessment and learning motivated by the expectations of others was strong in Japan, Malaysia, Sri Lanka, Nigeria and India, the corresponding relationship in England was much weaker, suggesting a stronger individual and weaker collective orientation among English students. Thus it would seem that the relationship between a motivation for learning based on selection and on task interest may not be universal, but contingent on context. Moreover, comparative propositions about learning and motivation, heavily influenced by a limited number of contexts, may overlook the importance of additional sources of motivation.

Have the symptoms of the diploma disease intensified?

A prediction of the general thesis is that countries which exhibited symptoms of the diploma disease in the 1970s when the thesis was developed would continue to do so, and that the symptoms would intensify over time. Sri Lanka was one of those countries. Have the symptoms of the diploma disease intensified in Sri Lanka? Have employment opportunities intensified the instrumental value attached by students to educational qualifications? Has the use of educational qualifications by employers intensified? Has the quality of the educational process become more examination-oriented?

Estimates of per-capita growth in the national economy suggest that a slow annual average growth of 1.3% between 1970 and 1977 was replaced by an annual average of 4% between 1977 and 1987. Recent estimates suggest rates of growth of gross national product of 8% between 1992 and 1993, and of 5% between 1993 and 1994. The annual growth of the labour force has declined slightly, from an annual growth of 2.5% between 1971 and 1980/1 to 2.2% between 1980/1 and 1990. At the same time the growth in female labour has been consistently higher than the growth of the male labour force, at 3% between 1971 and 1980/1 and 5% between 1980/1 and 1990 (Alailima, 1991). Employment opportunities abroad have increased over the 1970–90 period. Traditionally there had been a modest annual exodus (c. 1500) of highly skilled personnel to jobs in the countries of the old and new commonwealth. By the early 1990s it was estimated that around 100,000 people migrated annually for work to the countries of West Asia, among whom about half were young women seeking work as housemaids, many of them GCE O-level qualified (Nadaraja and Wijemanne, 1991). The growth in the economy may be attributed to its liberalization after the change of government in 1977 and the out-migration of labour to high growth West Asian economies during the 1980s. Both trends reflect the internationalization of the domestic economy and labour market.

The population has certainly become more educated and qualified. Between 1971 and 1986 the proportion of the population aged over five years with less than five years of primary education had declined from 24.5% to 11.6%, with the corresponding increases in the percentages of those with GCE O-level, A-level and university degree qualifications— from 3.7% to 10.8% for O level; from 0.9% to 2.1% for A level; and from 0.38% to 0.53% for university degrees.

Serious imbalances between the number of job seekers and job opportunities in the labour market remain, despite economic and educational growth. While open unemployment rates have declined, from 19% in 1971 to 15% in 1981 to 14% in 1990 (Alailima 1991), the composition of that imbalance has changed. Between 1970 and 1986 the unemployment rate among those aged 20–24 years of age who were GCE O-level qualified declined from 63% to 43%. Unemployment among the GCE A-level qualified almost doubled, from 27% to 52%, with an increase in absolute numbers from 9000 to 43,000. The bottleneck in education has shifted from access from the GCE O level to GCE A level to access from GCE A level to university. This is reflected in the extremely high percentages of students in GCE A-level classes in state-funded schools who follow private tuition classes in the afternoons and weekends. In 1989, survey data from across the country (excluding the war-torn north and east provinces) revealed that 92% of GCE A-level and 75% of GCE O-level students were following private tuition classes for an average of just over nine hours per week (de Silva *et al.*, 1991). Comparisons with earlier and more selective samples suggest that participation in private tuition has intensified over time (Little, 1994).

Whether schooling has become more or less examination-oriented since the 1970s is difficult to judge. The research findings of SLOG (1987), described earlier, suggested that the pattern of relationships between motivations for learning stimulated by selection and by task interest was no different in the mid-1980s in Sri Lanka and in England. However relationships between orientations for learning do not enable us to compare absolute levels of orientation, nor to make judgements about changes over time. Certainly the participation of the school-age population in examinations has increased, in proportion to the general growth of the education system. GCE O-level candidates increased from 300,000 in 1977 to 526,000 in 1991. GCE A-level candidates increased from 48,000 in 1975 to 160,000 in 1991.

Perhaps the most compelling evidence for the intensification of examination dominance comes from the aborted attempts to reform the assessment system. The dominance of the selection role of assessment in Sri Lanka, as compared with its certification, formative and monitoring roles, has thwarted several reforms designed to support a reorientation of curriculum content. Between 1972 and 1975 the content of the secondary school curriculum was reformed to make education more relevant to the lives of the majority of students. At the same time the GCE O- and A-level examinations were replaced by middle-secondary and upper-secondary examinations sat by students one year earlier. The examination reform was overturned in 1977. One of the major reasons for the rejection of the reform was the anxiety expressed by the vocal middle classes over access to higher education opportunities outside the country, providing some support for the dependency thesis about the interests of national élites lying in economic and educational systems outside national boundaries (Lewin and Little, 1984). A decade later an attempt was made to introduce continuous assessment in the public examination of O-level course content. Its life was to be even shorter than the reform of the mid-1970s, rejected by parents because of a mistrust of school-based assessment and by rural insurgents as discriminating against the poor rural child.

Ethnicity, language and political patronage

External commentators deplore the apparent grip of selective assessment on the educational experience of the Sri Lankan school student. Internal commentators too are increasingly concerned about the stress and anxiety experienced by young learners as they participate in the examination rat race (Economic Review, 1994). However, social and political changes in Sri Lanka appear to have attenuated the social construction of selective assessment as posing an educational problem. Issues of ethnicity and language and the growth of political patronage in job allocation appear to have reinforced the role of assessment as the most equitable means of allocating scarce resources in society.

Although the 'complicating factor of language' was recognized by Dore when comparing the history of educational development in Sri Lanka with that in Japan, the complication discussed was the tension between English, the language of the former colonizer, and the vernacular of the colonized in the immediate post-independence era. The further complication of there being two vernaculars, the Sinhala of the majority, and the Tamil of the minority, was not addressed. This, combined with a range of economic and political tensions between these two groups, has legitimated further the dominance of assessment's selective and allocative role in society.

Within a few years of independence Sinhala was declared the official language of government, except in Tamil-speaking areas in the north and east. Up until the mid-1950s the education of children from among the urban Tamil community in Jaffna was oriented to government jobs for which English was required. After the introduction of Sinhala as the official language opportunities for Tamils in government service declined. Tamil students, in response, focused their attention on higher education and science stream education in upper secondary school. This led, in turn, to a major ethnic imbalance in the representation of students in the prestigious science, engineering and medical faculties and a series of controversial changes in admissions policies to the university during the 1970s and early 1980s. Tamil-speaking children were and are still able to follow their entire education to the end of university in the Tamil medium.

In 1970 the United Front government introduced a system of admission to science-based courses in the university based on predetermined minimum marks according to the language medium of instruction; higher for Tamil students; lower for Sinhalese students. This change lasted one year. In 1971 GCE A-level marks were standardized by subject and by language in which the examination was sat, candidates were ranked by standardized marks and selected according to this rank order. This change was supported by those who attributed earlier differences in performance between Sinhalese and Tamil students to differences in facilities, marking or teaching. This would later be combined with district quotas. In 1974 the admissions policy changed again. Standardization was abandoned and replaced by a modified district quota system. Seventy percent of places were allocated on the basis of raw marks; 30% on a district basis. In 1976 the policy changed yet again. Standardized marks were reintroduced for the 70% of places allocated on order of merit. The remaining 30% were allocated on the basis of a district population quota including the educationally favoured districts of Jaffna and Colombo, excluded hitherto. Half the places in the district quota were reserved for students from educationally underprivileged districts. The policy changed again in 1979, in 1984 and 1987. By the early 1990s there were calls for all university places to be allocated on the basis of merit.

The issue of admission to higher education fuelled the tensions between the Sinhalese and Tamil communities and the eruption of conflict between the largely Sinhalese security forces and Tamil separatists in the late 1970s. The conflict escalated in the years following the anti-Tamil riots of 1983, subsided and escalated again with the arrival of the Indian peacekeeping force in 1987, achieved an uneasy truce with the new government of 1994, and continued full-scale early in 1995.

The frequency with which the admissions to university policy has changed and the intensity of political interest which has surrounded those changes are symptomatic of the conflict between the Sinhalese and Tamil communities over access to resources of various kinds. At the same time they have legitimated the role of examinations as the most equitable means of access to those resources. Examinations may have undesirable backwash effects on the processes of learning. But they also serve to legitimate inequalities of access to further education and employment opportunities.

A second major change in Sri Lankan society which has legitimated further the selective role of assessment has been the growth in political patronage though the 1980s. In the period up to 1977 public sector jobs were allocated on the basis of educational qualifications, experience and sometimes interviews. Politicians often provided additional written and verbal support for job applications. Such recommendations were part of an informal system which shadowed the institutionalized system of qualifications, experience and selection boards. However, after 1977, political recommendation was institutionalized through the 'job bank' scheme. Registrants with the scheme needed to meet minimum educational qualification criteria and to be recommended to it by a member of parliament. Political recommendation had thus been institutionalized and would became an important criterion of access to resources in many spheres of life, not just jobs. When the economy began to surge in the late 1970s and early 1980s this system of political 'chits' worked to the advantage of both the government politicians and their supporters. And with a long tradition of government elections every five years, and oscillations in political complexion, opposition supporters might have believed, not unreasonably, that their turn would come soon.

However, in 1982, President Jayewardene called a referendum to extend the life of parliament for a further five years. Young people, just too young to vote in 1977 would have to wait a full ten years before exercising their democratic vote for the first time. This led to a growing disaffection among youth, Sinhala and Tamil alike. Even the favoured government supporters began to perceive their access to resources constrained as government diverted more into its internal defence budget. The invitation by government of Indian peace-keepers to manage the Tamil crisis in the north led to an insurrection among disaffected Sinhalese youth in the south, to violence and counterviolence by the security forces in both the north and the south, to a state of emergency and a further delay of parliamentary elections to 1989.

A presidential commission on youth reported in 1990. Its recommendation to expand opportunities for youth in higher education was implemented in 1991 when eight provincial university affiliated colleges were established. The use of the job bank and political recommendations were abandoned in the early 1990s and the use of educational qualifications intensified. Interview information now supplements educational qualifications but interviews are used by employers for the sole purpose of establishing the authenticity of the submitted qualifications. Educational qualifications, combined with ethnic quotas, became the dominant means of securing jobs in the public sector in the 1990s.

Predictions based on comparative historical analyses of educational development in several countries led Dore to predict that the effects of the diploma disease would intensify in Sri Lanka. There are certainly signs, referred to earlier, that the intensity of examination orientation in the formal system of school education and the shadow system of private tutories has increased, and that examination success is being relied upon more heavily than before as the legitimate allocator of scarce resources. However the judgement of whether this dependence on examinations is being driven by or is reinforcing slow economic growth directly, must consider the likely effects on economic growth of resource allocation systems not based on educational qualification. Far from being perceived as a problem by most social groups in Sri Lanka, a dependence on examinations is seen as a solution to the competing claims of groups on scarce resources during a period of intense ethnic and political crisis.

In conclusion

International, comparative and historical studies of assessment have demonstrated some of the difficulties inherent in the attempt to establish universal propositions about assessment, motivation and learning. They also demonstrate the importance of a historical perspective on contemporary social arrangements. They establish the importance of understanding assessment *in context*, of exploring developments in relation to economic and cultural developments within and outside national boundaries, and of searching beyond the surface characteristics of assessment systems to their deeper significance for individuals and societies. As policy makers look beyond national boundaries for quick-fix solutions to domestic economic problems, they should pause awhile, and attempt to understand the context of assessment, before selecting those which might work 'back home'. Universal propositions about the antecedents and consequences of assessment for individuals, social groups, national and global societies may be attainable in principle but they will only carry conviction if they can embrace the diversity of assessment roles, histories and practices in a range of local and national contexts.

Nageswary and Ruth, both aged 11, will sit rather similar public examinations in a few years time. Nageswary's thoughts are already moving to them. Ruth's are not. Nageswary's parents pin their hopes on her success, believ-

ing that it will pave a way out of their present condition. If Nageswary suc-
ceeds she may be able to help others in her family to change their condition
too. Ruth's parents want her to do well at school but they will be happy
for her if she chooses an occupation which affords a reasonable standard
of living and job satisfaction and enables her to become independent. They
do not expect her educational success to make a major difference to their
own lives. Understanding how students and their families experience
assessment requires an understanding of assessment in context.

References

Alailima, P. (1991). *Labour Force—Current Situation*. Colombo: Marga Institute.
Allport, G. W. (1937). *Personality*. New York: Holt.
Azuma, H., Kashiwagi, K. and Ohno, H. (1987). Student learning orientations: the
    Japanese study. In SLOG, *Why do students learn? A six-country study of student
    motivation*. Institute of Development Studies Research Report Rr 17, Sussex.
Bowman, M. J. (1970). Mass elites on the threshold of the 1970s. *Comparative
    Education*, **6**(3), 141–60.
Cardoso, F. (1972). Dependency and development in Latin America. *New Left
    Review,* **74**.
Carnoy, M. (1974). *Education as Cultural Imperialism*. London: Longman.
de Silva, W. A., Gunawardene, C., Jayeweera, S., Perera, L., Rupasinghe, S. and Wijet-
    unge, S. (1991). *Extra-School Instruction, Social Equity and Educational Quality in
    Sri Lanka*. Report to the International Development Research Centre, Singapore.
Deci, E. C. (1975). *Intrinsic Motivation*. New York: Plenum Press.
Dore, R. P. (1976). *The Diploma Disease: Education, Qualification and Development*.
    London: George Allen and Unwin.
Dos Santos, T. (1973). The crisis of development theory and the problem of depen-
    dence in Latin America. In Bernstein, H. (ed.) *Underdevelopment and Develop-
    ment*. Harmondsworth: Penguin.
Economic Review (1994). Special Report on Private Tuition in Sri Lanka, Vol. 20,
    Nos 2/3. Colombo: People's Bank.
Featherstone, M. (ed.) (1990). *Global Culture*. London: Sage.
Frank, A. G. (1967). *Capitalism and Underdevelopment in Latin America*. New York:
    Monthly Review Press.
Galtung, J. (1980). *The True Worlds: A Transnational Perspective*. New York: Free Press.
GOSL (1972). *Interim Report of the Committee to Inquire into a Report on Public
    Examinations at Secondary School Level in Ceylon*. Colombo: Government of Sri
    Lanka, Government Printers.
Hextall, I. and Sarup, M. (1977). School knowledge, evaluation and alienation. In
    Young, M. and Whitty, G. (eds) *Society, State and Schooling*. London: Falmer Press.
Heyneman, S. and Loxley, W. (1983). The effect of primary school quality on
    academic achievement across 29 High and Low Income Countries. *American
    Journal of Sociology*, **88**, 1162–94.

Heyneman, S. and Ransom, A. (1990). Using examinations and testing to improve educational quality. *Educational Policy*, **4**(3), 177–92.

Lewin, K. M. (1984). Selection and curriculum reform. In Oxenham, J., *Education versus Qualifications?* London: Unwin Educational Books.

Lewin, K. M. and Little, A. W. (1984). Examination reform and educational change in Sri Lanka 1972–1982: modernisation or dependent underdevelopment? In Watson, K. *Dependence and Interdependence in Education*. London: Croom Helm.

Little, A. W. (1994). *The Diploma Disease in Sri Lanka: 1971–1993*. Research Working Paper No. 10. London: International Centre for Research on Assessment, Institute of Education.

Maehr, M. L. (1976). Continuing motivation: an analysis of a seldom considered educational outcome. *Review of Educational Research*, **46**(3), 443–62.

McNamara, V. (1980). School system structures, curriculum, SSCEP and functional motivation for learning. *Papua New Guinea Journal of Education*, **16**(1), 12–28.

Milner, M. (1972). *The Illusion of Equality, the Effects of Education on Opportunity, Inequality and Social Conflict*. San Francisco: Jossey-Bass.

Mortimore, J. and Mortimore, P. (1984). *Secondary School Examinations*. Bedford Way Papers 18. Institute of Education, University of London.

Mortimore, P., Sammons, P., Stoll, L., Lewis, D. and Ecob, R. (1988). *School Matters: The Junior Years*. London: Open Books.

Nadaraja, T. and Wijemanne, E. L. (1991). *Labour Force: Future Prospects and Employment Abroad*. Colombo: Marga Institute.

Nuttall, D. (1993). *Monitoring National Standards*. Research Working Paper No. 6. London: International Centre for Research on Assessment, Institute of Education.

Postlethwaite, T. N. (1987). Comparative educational achievement research: can it be improved? *Comparative Education Review*, **31**(1), 150–58.

Rahula, S. (1956). *History of Buddhism in Ceylon*. Colombo: M.D. Gunasena.

Riddell, A. (1989). An alternative approach to the study of school effectiveness in Third World countries. *Comparative Education Review*, **33**(4), 481–97.

Ruberu, R. (1962) *Education in Colonial Ceylon*. Kandy: Kandy Printers.

Rutter, M., Maughan, B., Mortimore, P. and Ouston, J. (1983). *Fifteen Thousand Hours: Secondary Schools and their effect on Children*. London: Open Books.

Sancheti, N. (1984). Institutional transfer and educational dependency: an Indian case study. In Watson, K. (ed.) *Dependence and Interdependence in Education*. London: Routledge.

Schiefelbein, E. and Simmons, J. (1981). *The Determinants of School Achievement: a Review of Research from Developing Countries*. Ottawa: IDRC.

SLOG (1987) *Why Do Students Learn? A Six Country Study of Student Motivation*. Institute of Development Students Research Reports Rr 17, Sussex.

Wallerstein, I. (1993). Comparative Sociology. In Outhwaite, W. and Bottomore, T. (eds) *Blackwell Dictionary of Twentieth Century Social Thought*. Oxford: Blackwell.

Waters, M. (1995). *Globalisation*. London: Routledge.

Weber, M. (1972). The Rationalisation of Education and Training. In Cosin, B. R. (ed.) *Education: Structure and Society*. Harmondsworth: Penguin Books and Open University Press.

CHAPTER TWO

Assessment and management: World Bank support for educational testing

Marlaine E. Lockheed

Over the past five years, World Bank support for monitoring the output of educational systems has grown, for two reasons: (a) greater country-level interest in monitoring learning achievement (WCEFA, 1990) and (b) greater donor interest in monitoring development impact (World Bank, 1994). These two interests coincide as education systems, worldwide, are becoming less centralized and the role of central education ministries redefined. Central education ministries are becoming less involved with day-to-day administrative issues and more involved with issues of policy, monitoring effectiveness and quality control. The consequence has been a sea change in perceptions about the purposes of educational testing and assessment, from a narrow focus on the selection and certification of individual accomplishment to a broad range of management purposes. This chapter describes how the demand for World Bank support for educational testing has grown over time and shifted commensurately with these broad international and donor changes.

International context for testing

Country level interest in monitoring learning achievement was expressed succinctly in the *World Declaration on Education for All* and *Framework for Action to Meet Basic Learning Needs* and, especially in Article 4, focusing on learning acquisition (cf. Foreword, passim).

One signal of this interest is the growing number of countries who are participating in cross-national comparative studies of educational achievement. Thirteen countries participated in the first International Association for the Evaluation of Achievement (IEA) mathematics study in 1964, 20 in the second IEA mathematics study in 1980, and over 40 countries are participating in the current (third) IEA mathematics study (IEA, 1993). Another is the inclusion of reading, mathematics and science achievement among 28 key education indicators for member countries of OECD (CERI, 1993).

The importance of monitoring the development impact of projects has been articulated for the World Bank in its overall framework for monitoring project performance. Bank guidelines indicate that every project supported by the bank must include not only a statement of objectives, but also specific benchmarks by which to judge progress in achieving these objectives (World Bank, 1994). Such objectives in the education sector include expanding access to education, reducing the cost per graduate of a given education cycle, and increasing the learning achievement of students. This leads directly to support for national efforts to monitor trends in learning achievement.

Decentralization of educational management to local political and administrative units, and—in some countries—to schools has also created a demand for information about education outputs, in three ways. First, managers at all levels (including school managers and classroom managers—i.e. principals and teachers) need information to improve the effectiveness of their decisions. Teachers need information about student performance to decide how to allocate their time to various elements of the curriculum—whether to continue to review fractions or to move on to decimals, for example. Principals need information about student performance to decide how to organize in-service training for teachers—whether to devote another training session to the topic of 'common problems student have with fractions' or to move on to another topic. Administrators need information about student performance to determine whether funds spent on a 'remedial fractions' programme has improved student comprehension of fractions. Parents need information about student performance to make judgements about school choice, when choice is legally possible. In all cases, information can help inform rational decision-making regarding the allocation of time and resources.

Second, decentralization of functions from central to more local agencies creates a 'function gap' at the centre which needs to be filled. Ministries of education are no longer required to make specific teacher appointments, select specific textbooks, sign leave slips and so forth. Resources previously used for these functions are freed for alternative uses. A common shift is from service provision to quality control, with the monitoring of quality—including the establishment of systems for the assessment of student learning—incorporated as a key element of control.

Third, the demand for information also arises from the need for greater accountability in good governance. Schools, school districts, local education authorities, and local governments are accountable for their technical and financial performance to parents, the public and higher levels of government and administration. The public needs to know whether the resources it provides, through taxes and school fees, are having the intended effect on student learning achievement.

To provide the information required by good management and accountability in education requires the systematic collection and use of information that supports efforts to improve school effectiveness. Such information includes: (a) indicators of how well the education system as a whole is doing with respect to achieving its qualitative and quantitative goals, (b) indicators of the performance of types of students, individual schools or groups of schools (such as school districts or geographic regions), and (c) indicators of the effectiveness of specific sets of policies adopted for school improvement. In all cases, an important set of indicators are those that seek to measure student learning achievement. However, not all measures of student learning achievement serve these purposes.

Purposes of testing

Measures of student learning achievement can be grouped into six general categories of 'tests', according to their purpose. I use the term 'test' to refer to 'any series of questions or exercises or other means of measuring the skill, knowledge, intelligence, capacities or aptitudes of an individual or group' (Anderson *et al.*, 1975). Educational achievement tests focus on skills, knowledge and capacities that are acquired through education (schooling); they do not seek to measure 'intelligence' or 'aptitude'.

The six most common purposes for educational achievement tests are:

- Selecting students for further education; examples are the Common Entrance Examination for secondary level education in Jamaica, the SAT and ACT college entrance examinations in the United States, or the Law School Admissions Test for graduate legal education, also in the United States.
- Certifying student achievement; examples are the O- and A-level examinations of the Cambridge Examination Syndicate or the Medical Board examinations to certify physician competency in many countries.
- Monitoring achievement trends over time; and example is the National Assessment of Educational Progress in the United States.
- Evaluating specific educational programmes or policies.

- Holding schools, regions, etc. accountable for student achievement; an example is the SIMCE in Chile.
- Diagnosing individual learning needs; examples include a wide range of tests for cognitive skills assessment.

While in some cases educational achievement tests designed for one of these purposes can be used for another one, this is more often the exception than the rule. In most instances, the use of tests designed specifically for one purpose (such as selection) for another purpose (such as monitoring achievement trends over time) is inappropriate for both technical and financial reasons.

Technically, the purpose of the test determines its design; different test designs are needed for different test purposes. Elements of design that would reflect purpose include:

- the difficulty level of the test questions
- the range of topics covered in the test
- the inclusiveness of the population tested
- the frequency of administration
- its format
- its standardization with respect to content, administration and scoring
- its reference to a norming group or performance criteria.

Tests designed for different purposes would differ on aspects of these design features (Table 2.1). For example, selection tests are designed to discriminate between individuals, and hence include many items of a difficulty known to discriminate between those who should be selected and those who should not; as a consequence, the range of competencies tested will be highly constrained. By comparison, a test designed to monitor national achievement is designed to assess a large number of competency areas but to provide little information about specific individuals. Similarly, tests designed to monitor trends over time are designed to repeat at least some items; tests designed only to select once need not be equated. Test misuse is a serious problem and confusion about multiple uses for tests is found everywhere.

Financially, using tests for more than one purpose can be not only inappropriate but also more costly. For example, using tests that have been designed for evaluative purposes (monitoring achievement or evaluating specific educational programmes or policies) as diagnostic instruments is not only unproductive but it will also increase costs (LeMahieu and Wallace, 1986). It is unproductive because evaluative testing instruments are not relevant, timely or brief enough to make for good diagnostic instruments.

Table 2.1 Management purposes of measuring learning achievement

	Monitoring progress toward national educational goals	Evaluating effectiveness and efficiency of specific policies	Holding schools accountable for performance	Selecting and/or certifying students student performance	Teacher assessment of individual
Example	National Assessment of Educational Progress, United States	Evaluation Analytique de l'enseignment Primaire, Benin	National Grade 7 Evaluation, Thailand	Certificate of Primary Education, Kenya	National Assessment, England and Wales
To whom administered	Sample of students in selected age or grade cohorts	Sample of students in sample of schools	Sample of students in sample of schools	All students in terminal year	All students in selected age cohorts
When administered	Periodically (annual, biannual)	One time	One time	Annual	Annual
Content objective	Selected domains (e.g. maths, science)	Selected domains	Selected domains	All domains of curriculum	Selected domains
Behavioural objective	Knowledge and higher order thinking skills	Knowledge and higher order thinking skills	Knowledge and higher order thinking skills	Knowledge and higher order thinking skills	Knowledge and higher order thinking skills
Format					
Objective	Yes	Yes	Yes	Yes	Yes
Performance	Yes	Yes	Yes	Yes	Yes
Standardized					
Content	Yes	Yes	Yes	Yes	Sometimes
Administration	Yes	Yes	Yes	Sometimes	Sometimes
Scoring	Yes	Yes	Yes	Sometimes	Sometimes
Reference					
Norming group	Yes	No	No	No	No
Performance criteria	Yes	No	Yes	No	Yes
Supplementary measures					
Student background	Limited	Yes	Limited	No	No
Classroom/school inputs	No	Yes	No	No	No
Classroom/school processes	No	Yes	No	No	No

It is more costly because using evaluative tests for diagnostic purposes implies testing every student in the population; thus, increasing costs of a test that only requires a good sample in order to obtain an accurate estimate of students' performance.

In addition, tests for purposes of selection and certification are typically 'high stakes' tests, where the outcomes have high value to the test taker. As a consequence, 'high stakes' tests have been shown to contribute to a wide range of distortions with respect to their impact (Shepard, 1991) and are generally considered to be inappropriate for providing an unbiased indicator of education system performance.

World Bank support for testing

World Bank support for educational testing has increased significantly in the recent past, catalysed by growing evidence regarding the poor performance of educational systems in developing countries, the desire by countries to manage their education systems better and more efficiently, the WCEFA's call for national systems for monitoring learning (WCEFA, 1990), and the need to monitor the impact of bank education lending operations on student learning. Projects that include support for testing activities are found in all regions and address a wide variety of purposes. The major type of test that is supported is tests for student selection or certification, although testing activities related to tests for monitoring achievement trends have become a focus of attention in projects in the most recent years.

The World Bank invests in educational testing through two major approaches: (a) education projects, such as free-standing primary or secondary education projects, and (b) education activities in non-education projects, such as social sector development projects. In both cases, testing activities are often subsumed under educational quality, institutional strengthening or management activities.

A review of educational testing in 85 bank education projects from FY75 (financial year 1975)—when the first testing subcomponent was supported—through to FY92 (which includes five projects under preparation for financial year 1993) found that the bank has supported testing for all six purposes for educational achievement listed above, at all levels, in all regions, and that support for testing increased over the nearly two decades reviewed (Larach and Lockheed, 1992).

Patterns

Bank projects support a variety of educational testing purposes. Thirty-three percent of projects that support testing activities support tests for student certification and selection, 21% support tests for monitoring student progress toward national educational goals, 19% support tests for evaluating the effectiveness of specific policies, 7% support tests for teacher diagnosis of student learning, 2% support tests for mixed or multiple purposes, while 17% of projects include more than one testing activity (Fig. 2.1).

Tests are supported at all levels of education (basic, secondary, tertiary) and projects frequently support tests at more than one level. Most monitoring, evaluation and diagnostic tests are prepared for basic education (96%, 86% and 93%, respectively). Selection/certification tests, in comparison, are more evenly supported across educational cycles, with 43% prepared for basic education, 40% for secondary, 38% for vocational, and 13% for higher education (Fig. 2.2).

Projects that support testing occur in all geographical regions, with thirty projects in Africa, twenty five in Asia, nineteen in Latin America and the Caribbean (LAC) and sixteen in Europe, the Middle East and North Africa (EMENA). Between regions, however, the percentage of projects that support testing differs sharply. Since FY75, a higher proportion of projects in LAC have included testing (33%) than have projects in other regions (25% of projects in Africa, 25% of projects in Asia and 23% of projects in EMENA). In part, this is due to the recent renewed lending for

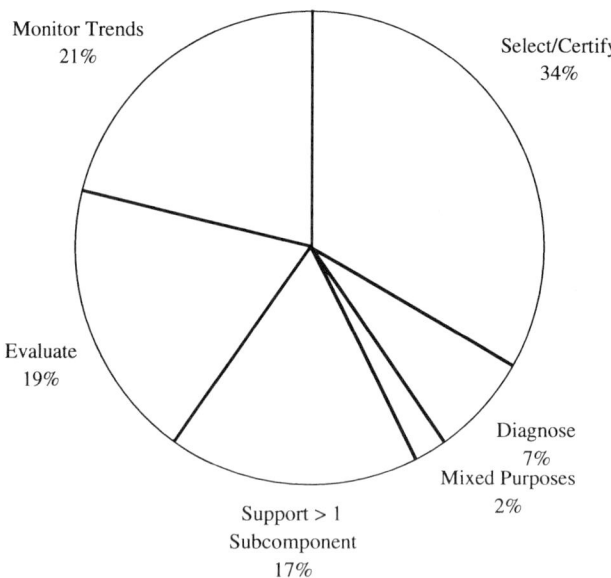

Figure 2.1 Percentage of projects providing support for six testing purposes.

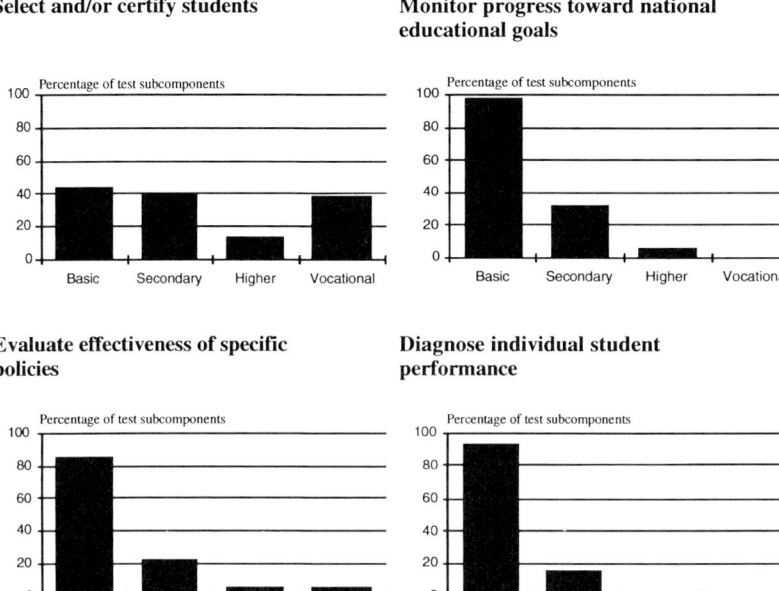

Select and/or certify students

Monitor progress toward national educational goals

Evaluate effectiveness of specific policies

Diagnose individual student performance

Figure 2.2   Projects with testing activities, by level of education.

education in Latin America, which coincided with the growth in lending for testing.

Trends

The number of projects with educational testing subcomponents, as well as the share of these projects in total education lending, has increased dramatically since the bank started lending for education. From FY63 to FY74, no projects supported testing subcomponents. In 1975, one project (5% of all education projects) supported these subcomponents. This rose to twelve projects in FY91 and ten projects in FY92, 46% and 38% of education projects, respectively, approximately doubling every five years (Fig. 2.3). This reflects the bank's increasing attention to monitoring the outcomes as well as the inputs of educational systems in developing countries.

Trends in the type of tests supported in bank projects are apparent (Fig. 2.4). The percentage of projects that support tests for selection and certification has been steadily growing over the past 15 years, while the

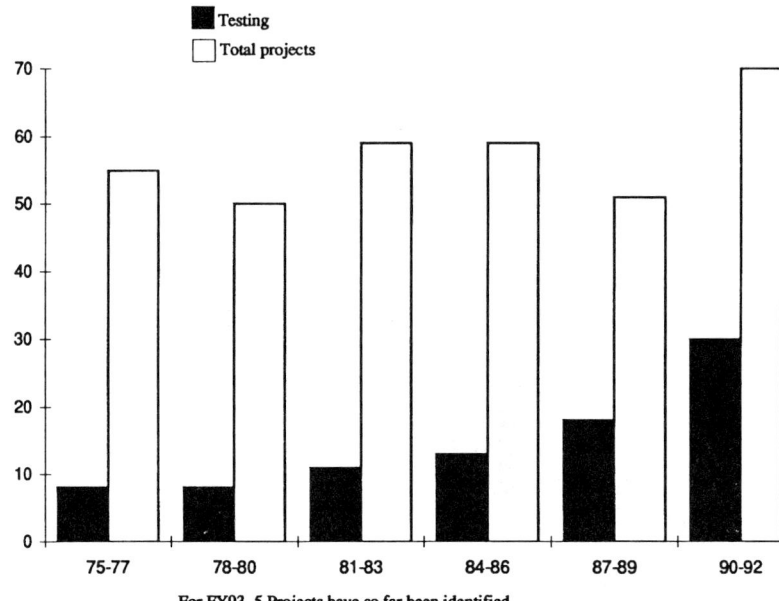

For FY93, 5 Projects have so far been identified

Figure 2.3   Projects with testing activities, by year.

percentage of projects that support tests for evaluating the effectiveness of specific policies has declined. Support for tests both for monitoring progress toward national education goals and for teacher diagnosis of student learning has increased sharply over the past decade. Monitoring has shown the greatest increase, growing from 0% of projects before FY88 to 27% of projects in FY91. The increase in monitoring reflects greater attention to strengthening the capacity of governments, particularly Ministries of Education, to gather education output information for purposes of policy analysis and improved management.

Regional differences in test purposes are also apparent. In Africa and Asia, nearly half (46% and 44%, respectively) of all projects support tests for selection and certification, while in the regions of EMENA and Latin America the largest share of project support is for tests for monitoring achievement trends (29% and 40%, respectively).

These regional differences reflect historical patterns of education in the region. African and Asian education systems have included national selection and certification examinations at all levels, and bank lending has been designed to strengthen existing examinations. Where national examinations have played a lesser role in educational systems, as in Latin America, bank

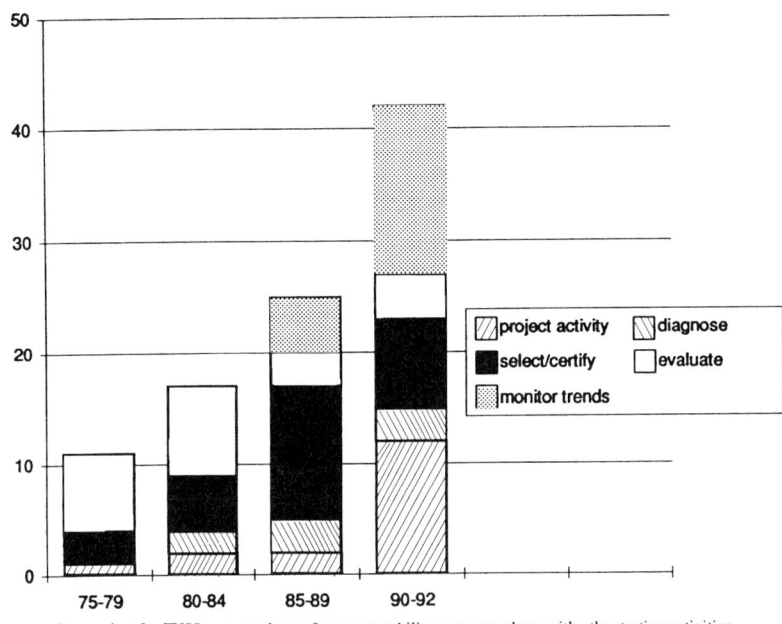

One project for FY90 supported tests for accountability purposes along with other testing activities.

Figure 2.4   Trends in support for five test purposes, FY75–92.

lending has been designed to support new systems of monitoring and assessment.

Institutional arrangements

In both developed and developing countries, most national programmes of testing for individual diagnosis, selection and certification purposes are operated by a specialized unit of the central ministry of education. By comparison, national assessments are treated differently by developed and developing countries. Developed countries with educational assessment programmes typically have located these programmes outside the government. Seventy-five percent of OECD countries that currently employ educational assessments house this programme in a single non-governmental institution contracted by the government to provide the assessment service. Typically these are independent research institutions (Brinkley *et al.*, 1991). An implicit rationale explains locating assessment activities outside the government. If the government is providing the service (that is, is running the schools), then it should not be its own auditor. The central government institution (typically the ministry of education) finances the national assessment and assumes responsibility for monitoring its implementation, but the actual technical work is assigned to a variety of appropriate non-governmental institutions.

In contrast to OECD countries, most developing countries with national assessment systems house them in central education ministries. For example, most countries that have requested World Bank support for assessment systems have proposed to house them *within* the government. Even where assessment systems have previously been housed outside the government, such as in Costa Rica and Chile, the assessment systems will be gradually transferred to government units. Only two countries—Haiti and the Philippines—have proposed to subcontract various technical aspects to non-government institutions, and major responsibilities for providing the assessment service are expected to remain with the government (Horn *et al.*, 1991; Lapointe, 1990; Morales Frias, 1990.)

Ministries of education in developing countries have used two strategies for institutionalizing the assessment system. The first is the 'single unit' strategy in which a single unit is responsible for the development and management of the assessment; of 25 projects with assessment components for which information was available, 14 (56%) use this strategy. The 'single unit' strategy is implemented in two different ways, by: (a) creating a new division and/or unit within the ministry, and (b) expanding the responsibilities of an already existing unit. New units are proposed in six (43%) of the 14 projects using the 'single unit' strategy, while institutional responsibilities are being expanded in the remaining eight (57%). For example, Mexico assigns responsibility for assessment to the Ministry of Education's evaluation unit, Direccion General de Evaluacion y de Incorporacion y Revalidacion. (Secretaria de Educacion Publica de Mexico, 1991).

A second strategy for integrating the assessment system is the 'collaboration' strategy in which one government unit coordinates collaborative efforts of various divisions and/or units within the ministries of education. Seven (28%) of 25 projects proposed to use this strategy. The project in the Philippines is an example of this second institutionalization strategy (Lapointe, 1990). With support from the World Bank, the Philippines Department of Education has established a student assessment system. The principal responsibility for the operation of the system lies in the already established National Educational Testing and Research Center (NETRC), whose director has enlisted the cooperation and contracted services of various bureaux and agencies, including the Center for Educational Measurement. The NETRC is advised by two committees: a national advisory committee on policy matters and a technical advisory committee on design and implementation. The national advisory committee includes educators, academics, university administrators, business leaders, representatives from small business and organized labour, regional and national political leaders, teachers and parent representatives.

Institution strengthening

One consequence of central government control of assessment systems is the tendency for ministries of education to request donor support for institutional strengthening related to assessments. About half of bank-supported projects provide support for strengthening four key aspects of the institutional base for assessments: institutional stability, staff quality, local capacity for training and professional communication.

An example comes from a project for Turkey. With support from the World Bank, Turkey has strengthened its institutions responsible for testing and conducting research (Ministry of National Education, Turkey, 1989). Specifically, the Ministry of Education has strengthened and expanded the role of the Testing and Evaluation Centre by transforming it into a Testing and Evaluation Division (TED) of the Education Research and Development Centre (EDRC). The new assessment functions of the TED are clearly delineated as: (a) developing standardized tests for selected grades and subjects; (b) analysing tests results to improve the curricula and classroom pedagogy; (c) undertaking policy research studies on the impact of education inputs on student achievement; (d) participating in a comparative assessment of the achievement levels of school children in Turkey and those in other countries; (e) experimenting with different testing formats; and (f) monitoring changes in curricula, instructional materials, and school practices and revise tests accordingly. Staff stability has been encouraged by emphasizing professional and high-quality management for TED staff and by developing a career structure. The TED employs highly qualified staff with PhDs in testing and evaluation, psychometrics, and statistical research methods who are supported with MA graduates in the necessary fields. Staff are encouraged to pursue higher educational qualifications and to produce and present papers at national and international conferences and symposia. A significant information/library back up service on testing, validation, evaluation developments and related subjects has also been provided.

Conclusions

The World Bank's support for educational testing has grown significantly over the past two decades, and has shifted from providing support principally for selection and certification tests to providing support for a wide range of test purposes, with the sharpest increase in support for national assessment systems. It is too soon to determine the impact of these programmes on educational systems and their management in all countries with assessment programmes. However, benefits from national educational assessments can already be observed. For example, in Chile, the results of the national assessment have been combined with other social indicators to assist the national Ministry of Education in targeting additional support for the poorest schools (Himmel, 1992). Each potential school is reviewed according to the average student performance on the SIMCE (national assessment) test, socioeconomic level, degree of rurality, and number of primary grades offered. The achievement test score accounts for 50% of the school score. On the basis of the score, schools are rated as 'high risk', 'medium risk' and 'low risk'. Forty-six percent of the available resources for school improvement are targeted at 'high risk' schools and another 46 percent are targeted at 'medium risk' schools. Within risk categories, schools compete for funds by proposing school improvement activities for support. The 1994 school-level scores on SIMCE will have provided evidence regarding the effectiveness of four different educational interventions which will be introduced on a pilot basis before widespread introduction; interventions that do not boost learning will not be eligible for a larger-scale implementation. (See also Chapter 7.)

In many countries, national assessment systems are enabling ministries of education to monitor their own progress, evaluate the potential impact of experimental programmes and their cost-effectiveness, and improve the quality of their educational planning. Information from national assessments can inform teaching and learning processes, when this information is disseminated broadly. The World Bank, as well as other donors, is assisting many countries in strengthening the institutions responsible for national assessment systems. This strengthened capacity to monitor will enable the bank and other donor agencies to monitor the effect of their support on an important development goal: the learning achievement of children.

References

Anderson, S., Ball, S. and Murphy, R. (eds) (1975). *Encyclopedia of Educational Evaluation*. San Francisco: Jossey-Bass.

Brinkley, M., Guthrie, J. W. and Wyatt, T. J. (1991). *A Survey of National Assessment and Examination Practices in OECD Countries*. Switzerland: Lugano.

CERI (Centre for Educational Research and Innovation) (1993). *Education at a Glance: OECD Indicators*. Paris: OECD.

Himmel, E. (1992). A case study of the use of national assessment in Chile. Santiago, Chile: Processed.

Horn, R., Wolfe, L. and Velez, E. (1991). *Developing Educational Assessment Systems in Latin America: A Review of Issues and Recent Experience.* Latin America and the Caribbean Technical Department, Regional Studies Program, Report No. 9. Washington, DC: World Bank.

IEA (1993). *IEA Guidebook: Activities, Institutions and People.* The Hague: International Association for the Assessment of Educational Achievement.

Lapointe, A. (1990, Draft). *The School Assessment System.* Background paper for Philippines Secondary Elementary Education Project. Washington, DC: World Bank.

Larach, L. and Lockheed, M. E. (1992). *World Bank Lending for Educational Testing.* Population and Human Resources Department, PHREE Background Paper 92/62R, The World Bank.

LeMahieu, P. G. and Wallace Jr., R. C. (1986). Up against the wall: psychometrics meets praxis. *Educational Measurement: Issues and Practice.* Spring, 12–16.

Ministry of National Education, Turkey (1989). *National Education Project Phase-1.* Republic of Turkey, Project Implementation Unit, Turkey: Processed.

Morales Frias, J. (1990). *Chile: Sistema de Medicion de la Calidad de la Educacion (SIMCE).* World Bank, Latin America Technical Department. Washington, DC: Processed.

Secretaria de Educacion Publica de Mexico (1991). *Proposal for a Quality Assessment System for Primary Education in the states of Chiapas, Guerrero, Hidalgo and Oaxaca (Proyecto de evaluacion de la calidad de la educacion primaria en los estados de Chiapas, Guerrero, Hidalgo y Oaxaca).* Mexico, DF: Processed.

Shepard, L. A. (1991). *Will National Tests Improve Student Learning?* Paper presented at the American Educational Research Association Forum, Washington, DC, June.

World Conference on Education for All (WCEFA) (1990). *Meeting Basic Learning Needs: A Vision for the 1990's.* New York: Inter-Agency Commission for World Conference on Education-for-All.

World Bank (1994). *The World Bank Group: Learning from the Past, Embracing the Future.* Washington, DC.

CHAPTER THREE

After Jomtien: UNESCO's current policy on assessment

Vinayagum Chinapah

Current policy, research and programme on assessment

Since the World Conference on Education for All (WCEFA), held in Jomtien, Thailand in March 1990, all the international partners involved (UNESCO, UNICEF, UNDP and the World Bank) have carried out a number of important follow-up programmes, projects and activities aimed at assisting member states in the implementation of the goals of Education-for-All (EFA).

The mandate

Against this background, there has been a call for quality improvement in basic education: making the learning environment conducive so that children actually learn how to improve their lives. Quantitative base-line information is no longer enough. Educational authorities now want to know *how much is actually being learned.* But how do educational authorities monitor what pupils learn? What are the factors influencing learning achievement that should be examined? How can national capacity be developed to monitor learning on a permanent and sustainable basis?

Basic education policy

UNESCO's current basic education policy is targeted towards programmes of expanding access and improving quality and relevance in education. The main objectives are:

- to promote access to primary education or non-formal and non-traditional education alternatives, as a complement to classroom teaching, for all children, with special emphasis on girls and those more difficult to reach; and

• to contribute to the overall improvement of the quality of basic education with a view to increasing pupils' levels of learning achievement and reducing drop-outs and repetitions.

Several projects aimed at improving quality and learning achievement, as identified by the International Symposium on Strategies on Ensuring Success in Primary Schooling, have been launched in collaboration with donor agencies, member states and non-governmental organizations (NGOs) (UNESCO, 1992). UNESCO supplies the technical support for national projects, especially in the regions of Africa, Latin America and the Caribbean. These include programmes aimed at school success and retention, the dissemination of information on practical experiences and successes, the improvement of school success for children belonging to cultural minorities, and assistance in the revision of primary school curricula to meet local learning needs.

Research and programme on assessment

UNESCO's programme emphasizes the eradication of illiteracy by the year 2000 and the Jomtien framework for action to meet basic learning needs. Regional and subregional programmes have been designed to revise teaching methods and assessment tools, emphasizing the child's learning process and support provided by the family and the community.

UNESCO's International Institute for Educational Planning (IIEP) has launched an interregional research programme to strengthen national capacities of member states for planning and monitoring quality improvements in basic education services, with financial assistance from UNDP (Carron, 1992). This programme conducted in-depth analyses of the teaching/learning conditions of primary schools in different contexts (Guinea, India, China and Mexico) to determine the extent to which these schools equip learners with basic education, and the factors associated with differences in school results.

Another research programme, implemented by the UNESCO Institute for Education in Hamburg, is developing an innovative methodology for monitoring and evaluating non-formal basic level education for out-of-school children, young people and adults. A series of interregional orientation seminars and training workshops has been organized in order to design monitoring systems and appropriate evaluation techniques and procedures.

The joint UNESCO–UNICEF international project, *Monitoring Education-For-All Goals: Focusing on Learning Achievement*, was designed to answer the need for a new approach to educational assessment. Initiated in 1992, its main objective—to improve the knowledge base concerning the various factors affecting learning in basic education—focuses on building national capacities for monitoring. The primary objective is to assist countries in context-relevant monitoring. Unlike traditional forms of evaluation, based largely on rigid, normative and standardized procedures, this project stresses flexibility, openness and the opportunity to correct weaknesses at every step.

Monitoring learning achievement: importance and implications for educational assessment

The multiple functions of educational assessment

Educational assessment generally serves one of three purposes: (a) providing baseline information in order to ensure appropriate corrective measures; (b) assisting in selection/certification; and (c) influencing policy-making. Although the most common purpose of educational assessment has been to select students for further studies or to certify them for entry into the job market, still there is an urgent call for multipurpose educational assessment.

A similar trend occurred in the field of educational evaluation using sampling survey techniques. Monitoring learning achievement is gaining momentum at national level through capacity-building modalities while the conventional approach based upon large-scale cross-sectional surveys is becoming more and more questionable in terms of its importance for national decision-making and for immediate feedback to bring about quality improvement in education. The joint UNESCO–UNICEF project follows strictly the monitoring approach as requested by the member states with important modalities and strategies to promote national capacity-building in the field of educational assessment.

We have already seen in Chapters 1 and 2 that educational assessment serves a range of purposes, including the monitoring of learning achievement. One approach to monitoring has employed large-scale cross-sectional surveys and the generation of cross-national league tables. A pioneering example of this approach came from the IEA. However, as we shall see in Chapter 4, there are many methodological problems associated with these surveys. Moreover, the learning problems faced by millions of students in the basic education cycles of developing countries were not addressed adequately by these surveys.

The approach of the UNESCO–UNICEF monitoring project addresses these more fundamental learning problems. It focuses on the contexts, needs and priorities of specific countries. It emphasizes monitoring and capacity-building, and aims to improve the quality of the education offered by identifying strengths, weaknesses and trends in the systems, from the school to the country level.

Identifying strengths and weaknesses in the system is the first step; being able to act on them is the next, and most important step. And, in order to best do so, the monitoring process must aim for a sustainable, ongoing, built-in mechanism, which permits and encourages constant improvement. Moreover, the monitoring process must also be flexible and easily adaptable to each country's context.

The joint UNESCO–UNICEF monitoring project

The UNESCO–UNICEF project, *Monitoring Education-For-All Goals: Focusing on Learning Achievement*, is an important part of the WCEFA follow-up activities. Five countries, China, Jordan, Mali, Mauritius and Morocco, each with different development characteristics and experience in conducting educational survey studies, were chosen as pilot countries.

By the end of first phase of the project (1992–95), national teams in at least 20 countries will be fully trained, self-sufficient and ready to undertake the monitoring function on a permanent basis. The five pilot countries will play a key role as models and resources for others in the respective regions, and will also be the natural hosts for regional workshops and meetings. As the multiplier effect takes over, UNESCO and UNICEF's direct roles will gradually diminish.

In cooperation with its Jomtien partners, the monitoring project focuses its activities on the development of regional and national indicator systems for identifying the various factors affecting learning achievement and analysing the impact of these factors on participation, school wastage and learning achievement. The project aims to develop national capacities in monitoring. Monitoring is seen as a continuous activity, and the national task forces are expected to become self-sustaining after approximately three years. The project focuses upon the training of a selected group of people to carry out the project during its first phase. This group will subsequently train a wider group to monitor basic education programmes on a continuous basis.

The project focuses on primary education as the cutting edge of any EFA strategy. Working links are being maintained with complementary projects which may focus on the monitoring of non-formal basic education and literacy programmes. The strategic goal of the project is to ensure the appropriate linkages for actions at different levels of policy making and implementation of EFA goals.

The key approach is one of school-based sample surveys conducted in each cooperating country on a yearly or two-yearly basis. A simple, flexible and sustainable survey method is proposed in order to provide policy makers and front-line implementors with a monitoring information-base for the quality improvement of basic education programmes and projects. The nature and scope of the method is determined by participating countries themselves.

The national task forces

The success of this project resides largely in the commitment to and active participation in the process of developing and/or strengthening national capacities for monitoring of basic education programmes. The establishment of a national task force to participate in the design and implementation of the project is, therefore, a crucial first step, followed by intensive training efforts.

The national task force coordinates all project activities, working independently of but closely with UNICEF and UNESCO, and is responsible for the design and proper administration of the surveys. It oversees the collection, analysis and reporting of data and is responsible for project budgeting and accountability. An important part of the task force's role is the selection and training of local personnel to participate in the data collection and analysis throughout the country.

Selection of the national task force is the first and perhaps most important step in national capacity-building, as the ultimate aim is for each country's monitoring system to be self-sufficient after three years. In China, for example, the national task force comprises (i) the Department of Basic Education of the State Education Commission; (ii) the China National Institute for Educational Research; (iii) the Capital Teacher Training College; (iv) the Curriculum and Teaching Material Research Institution of the People's Education Press; and (v) the Provincial and County Level Bureaux and Units for Educational Research.

Building national capacities

The objectives set for capacity-building programmes can only be realized within the framework of human resources development (training of a group of national trainers) and infrastructure building (development of a national monitoring environment) of a permanent nature. In some countries, skilled human resources are present but not always adequately identified and assisted. Infrastructure building has often consisted of one-shot external inputs with no adequate national involvement and support for its sustainability. Table 3.1 shows the number of national workshops, core trainers, subnational workshops and peripheral trainees planned during the first phase of the project.

Table 3.1 National capacity-building: modalities and outputs

	Number of national workshops	Number of core trainers*	Number of subnational workshops	Number of peripheral trainees†
China	7	114	22	6645
Jordan	5	53	18	195
Mali	3	12	3	64
Mauritius	3	38	8	52
Morocco	8	37	9	103

*Core trainers—policy-makers, front-line implementors, experts and specialists.
†Peripheral trainees—regional, subregional officers, inspectors, area supervisors and teachers.

Development of indicators for educational assessment

The indicators of educational assessment relate to basic learning competencies (BLCs)—literacy, numeracy and life skills—of students. They also reflect factors found to influence learning achievement and educational access and equity. Thus, the indicators cover three domains:

1. *Learning achievement*: the BLCs of literacy, numeracy and life skills.
2. *Factors influencing learning achievement*: student characteristics and selected home background factors; the school setting and selected teaching–learning factors; and the community environment and selected school/community-related factors.
3. *Access and equity*: gender-specific enrolment, admission and participation rates, drop-outs and repeaters; attendance rates; educational disparities; out-of-school children, enrolment of disabled children, etc.

For the purpose of this chapter, we will distinguish: (a) the measurement of the 'common core' of basic competencies—in literacy, numeracy and life-skills—as measured by tests; and (b) the collection of reliable information concerning key scholastic and non-scholastic factors, through questionnaires.

It is important to note that this is the first time that the mastery of 'life skills' has been considered to be as important as literacy or numeracy acquisition in an international project. 'Life skills' refers to problem solving and social and 'attitudinal' skills in areas ranging from health and nutrition to road safety to civic responsibility. Naturally these vary from culture to culture and from country to country.

BLCs—common and specific items

Before beginning to develop the tests, the country must clearly define the skills it wishes to measure in order to guide the construction of the questions. Only then will the results of the tests have any meaning, in the particular context of that country. The BLC approach was developed in the International Workshop (UNESCO, 1993) by participants from the five countries of the project and from the inputs of the experts and consultants at this workshop. While the five pilot countries were obviously inspired by the results from the International Workshop, in each case they made an effort to develop the questions in accordance with their national context (UNESCO, 1994).

Mauritius, for example, defined BLCs of each domain in the following way.

Literacy: A nine-year-old child should be able to use reading and writing skills to meet the needs of everyday life and to sustain further learning. He or she should therefore be able to read and understand a variety of texts (functional and imaginative) appropriate for his or her age level and write legibly, clearly and correctly, words, phrases and sentences to convey required information.

Numeracy: A nine-year-old child is numerate if he or she is able to: (i) read and write numbers both in words and in figures (up to four digits); (ii) perform the four fundamental operations of arithmetic on numbers; (iii) solve simple shopping problems; (iv) read and write time on the 12-hour and the 24-hour clocks and read calendars; (v) compare lengths and convert units of measurement of length, mass and time into related units; and (vi) identify simple geometrical shapes and appreciate their properties.

Life Skills: A nine-year-old should have such basic knowledge, understanding, skills, attitudes and values that will enable him or her to cope with the demands of different life situations and which the educational system would make on him or her.

Table 3.2 shows the number of common and specific items in the domains of literacy, numeracy and life skills found in each country. Common items are those found in at least two of the five countries while specific items are unique to one country.

An analysis of the tests used across the five countries in the domain of life skills illustrates the design of country-specific instruments within an international project.

Table 3.2   Common and specific achievement test items in the five countries

Questionnaire items	China*	Jordan*	Mali*	Mauritius	Morocco
Literacy					
No. common items	10	22	11	7	23
No. specific items	10	14	5	6	7
All items	20	36	16	13	30
Numeracy					
No. common items	99	39	11	12	10
No. specific items	21	11	2	1	5
All items	120	50	13	13	15
Life skills					
No. common items	20	30	18	11	23
No. specific items	24	20	2	14	1
All items	44	50	20	25	24
Grand total					
No. common items	129	91	40	30	56
No. specific items	55	45	9	21	13
All items	184	136	49	51	69

*China (Grades 4 and 6); Jordan (Grades 4 and 8); Mali (Grade 5 and literacy classes).

At the International Workshop the five pilot countries agreed to include test items on four knowledge themes:

• health (disease, hygiene, nutrition, reproduction)
• everyday life (accidents, safety, home life)
• environment (conservation, civics, general knowledge)
• working life (crop cultivation and harvest, work and income-generating habits)

Health. All countries included health questions relating to issues such as vaccination, symptoms of common illnesses, first aid and medicines, as well as questions about smoking, washing hands, and brushing teeth. Jordan was especially interested in diseases transmitted by insects, and Mali in the subject of malaria. Several questions on the Chinese test concerned study habits and eyesight (e.g. how far away from your eyes to hold a book). Water was a frequent theme, as well as the importance of milk (its importance to one's health as well as how it should be stored) and the importance of well-balanced meals and nutritious foods. China was interested in technical aspects of food preparation (cooking, refrigeration) and Jordan asked about the harmfulness of certain foods and beverages (including tea and coffee), eating properly and whether it was preferable to breast-feed.

Everyday life. A wide range of issues were covered under 'accidents', especially day-to-day domestic accidents such as nosebleeds, cut fingers, snakebites and dangerous liquids in the eye. In Mauritius questions were asked about reactions to danger, such as what to do if you see a child drowning or if one of your team is hurt playing ball, what to do when caught in a thunderstorm, or how to tell the difference between unlabelled bottles. Each country asked about road safety, and most asked personal safety questions such as how to deal with strangers. Jordan was the only country to ask about the consequences of large families. Mauritius asked about appropriate clothing for different seasons, and China asked some very practical questions such as what to do if you lose your bicycle or how to address an envelope.

Environment. The most frequent questions about the environment concerned protecting forests and trees (including how to cook using as little wood as possible, preventing forest fires, caring for young trees). Mali asked about water pollution and Mauritius dealt with a number of environmental topics such as the protection of flowers and birds' eggs and the importance of keeping public places clean. Every country asked who was the head of state, and the names of the neighbouring countries. Mali asked about the colours of the national flag and Jordan inquired about the national money and the date of independence.

As far as 'civic life' is concerned, all five asked something about social responsibility—for example, what to do if you find something that doesn't belong to you. Questions on the Chinese and Mauritius tests dealt with relations with other people—what to do if a neighbour's radio is bothering you, or about a schoolmate who stutters. In the 'general knowledge' category, questions included the boiling point of water and what kind of book to consult if unable to spell a word.

Working life. Questions about harvests and cattle appeared on all the tests, as well as which insects can be harmful to crops, or what are the consequences of a dry year. Jordan seemed to be especially concerned by such questions. In Mauritius, children were tested on skills related to working the land. In China, children were asked about sewing, woodworking, tree transplantation and electricity. In Mali, children's agricultural skills were examined.

Key learning environmental factors

The overall conceptual framework of the monitoring project is built upon the conviction that different forms of assessment (e.g. teacher-made tests, examinations) have specific functions, and that differences in performance on these assessments are often due to contextual, personal, behavioural and attitudinal factors. The monitoring project provides a conceptual framework where these factors can be identified and analysed in view of their importance in order to design corrective measures and to improve the learning outcomes of children at critical moments of their schooling. Countries participating in the monitoring project have developed, pretested and administered questionnaires for pupils, parents, class teachers and school heads which address those background variables and other factors present in the home and school environments which may influence school achievement results. (For further details see UNESCO, 1994.)

Common and specific questionnaire items

Table 3.3 overleaf presents a meta-analysis of the different sets of questionnaires with country common and specific questions.

Preliminary report of country findings

The national assessment of BLCs

If the mandate of the project is to facilitate countries in designing an appropriate monitoring system for learning achievement, it is equally important to assess the extent to which the target set for the 1990s at the WCEFA, Jomtien, 1990, namely 'improvement in learning achievement such that an agreed percentage of an appropriate age cohort attains or surpasses a defined level of necessary learning achievement', is being reached. It is only possible to do so through broad-based consultations on basic learning competency targets, as defined by the national task force of each participating country and where criterion-reference testing is being applied to set these targets. This has already been the case in the first five countries of the monitoring project and the results are presented in Table 3.4. In all five countries and in most BLC domains, the target of 80% attaining or surpassing a defined level of necessary learning achievement is being met.

Table 3.3 Meta-analysis of questionnaire items in the five countries

Questionnaire types	China	Jordan	Mali	Mauritius	Morocco
Pupil					
No. common items	15	20	11	16	15
No. specific items	0	0	1	1	1
Class teacher					
No. common items	15	7	11	10	14
No. specific items	7	0	1	2	6
School					
No. common items	37	36	23	10	28
No. specific items	1	3	0	0	6
Parent					
No. common items	13	15	8	26	12
No. specific items	0	3	1	6	1
Total					
No. common items	80	71	52	62	68
No. specific items	8	6	3	9	14

Note. Common items are those found in at least two of the five countries while specific items occur only in a given country.

The majority of children in China in Grades 4 and 6 master the BLCs in all three domains, although in the life skills domain there is still room for improvement. In Jordan, more than 80% of the children have mastered what they ought to learn. In Mauritius, although there is a significant proportion of Grade 4 pupils who master the BLCs, there is a marked difference in levels of mastery of the two languages, i.e. English language (83%) compared with French language (70%). The performance in numeracy is also lower than that in other domains. In Mali, the target of 80% is being met in all three domains. It is important to note that the performance in the life-skills domain is higher than that in the other two domains in Mali. A majority of Moroccan children have mastered BLCs in all three domains, although the performance is lower in the domain of numeracy.

Within-country findings

Within-country analysis is as important as cross-country analysis, if not more so. In this part, some selected findings on BLC domains between

Table 3.4 Profiles of basic learning competencies (BLCs) by domains at Grade 4 level

Basic learning competencies (BLCs by domains)	China	Jordan	Mauritius		Mali	Morocco
Literacy			*			
Grade 4			(i)	(ii)	†	
Percentage BLCs and above	89	82	83	87	81	83
Percentage below BLCs	11	18	17	30	19	17
Numeracy Grade 4						
Percentage BLCs and above	94	83		70	83	80
Percentage below BLCs	16	17		30	17	20
Life skills Grade 4						
Percentage BLCs and above	72	80		82	89	84
Percentage below BLCs	28	20		18	11	16

*In Mauritius, BLC results in literacy are in two languages: (i) English and (ii) French.
†In Mali, BLC results are for pupils at the beginning of Grade 5.

urban and rural schools, between boys and girls, and between private and public schools are presented for the five countries.

The results are shown in Table 3.5 and the following trends can be observed.

Urban–rural differences. Students in rural schools have traditionally done worse on achievement tests throughout the world, for many reasons, including lower parental income and educational levels, poorer facilities, more poorly trained teachers and lower expectations.

In China, as in Mali, the urban–rural differences are not as marked as in the other three countries. In both countries urban school children do slightly better than rural school children in all three BLC domains, although the differences are non-significant in the life-skills domain in China.

Table 3.5 Within-country differences by region, gender and school type in basic learning competencies (BLCs) at Grade 4 level

Basic learning competencies (BLCs by domains)	China	Jordan	Mauritius †		Mali ‡	Morocco
			(i)	(ii)		
Literacy						
Grade 4						
Mean point-score						
differences						
Urban vs. rural	2.43	7.29	6.8	14.5	3.60	18.36
Girls vs. boys	*	6.71	5.1	4.4	−0.13	7.24
Private vs. public	−0.56	20.94	−6.8	−7.4	4.19	19.25
Numeracy						
Grade 4						
Mean point-score differences						
Urban vs. rural	2.94	4.68	10.5		2.34	13.63
Girls vs. boys	*	1.46	0.9		−1.66	1.45
Private vs. public	1.15	14.62	−13.0		4.15	19.84
Life skills						
Grade 4						
Mean point-score differences						
Urban vs. rural	0.54	2.0	7.3		3.21	15.36
Girls vs. boys	0.60	10.0	2.4		0.92	4.62
Private vs. public	0.49	7.0	−10.4		−0.37	13.15

*Not analysed.
†In Mauritius, BLC results in literacy are in two languages: (i) English and (ii) French.
‡In Mali, BLC results are for pupils at the beginning of Grade 5.

In Jordan, significant differences are noted between urban and rural achievement in literacy and numeracy, while in Mauritius and Morocco marked regional differences are observed in all three domains, with urban children doing far better than rural children.

Gender differences. In almost every country, the mean test scores were equal or higher for girls than boys. Whatever the reasons, throughout the world, the changing role of women in society has made the question of success or failure of girls in schools an increasingly important policy issue.

Table 3.5 confirms this trend. Girls usually outperform boys significantly in languages while female performance in numeracy is slightly higher but non-significant. In Mali, however, it is interesting to note that only in the life-skills domain does this trend persist. In Jordan, Mauritius and Morocco girls outperform boys in all BLC domains. The gender difference varies from one domain to another, being much greater in literacy, somewhat higher in life skills and non-significant in numeracy. These results clearly confirm those found in several studies in other developing countries. Once girls get into schools and equal access is ensured, they tend to outperform boys at basic education level. As they stay longer in the system, however, these differences tend to diminish (Eshiwani, 1983; Chinapah, 1983; Biazen and Junge, 1988; Elley, 1992; Postlethwaite and Ross, 1992).

Private and public schools. Private and public schools differ from country to country along a continuum. In some countries private schools are financed by the state or run by private bodies (e.g. religious authorities) and in others they are subsidized by private enterprises or school fees. Public schools are not always entirely free nor always run by the state. Some are community-based schools, self-help schools, and so on. It is, therefore, not entirely clear what constitutes a 'private' or a 'public' school and the subsequent analysis using these terms must be interpreted with caution. At the basic education level in most countries, however, private schools tend to draw children from families with relatively higher socioeconomic levels from urban or semiurban settings. Private schooling has traditionally produced high-achieving students, for much the same reasons as given above for urban schools. But research has also indicated that students from wealthier, highly-educated families would do well in school regardless of the institution they attended.

Although there are no significant differences between privately-run and publicly-run schools in China, children from public schools do better than children from private schools in the literacy domain. Performance in the life-skills domain is somewhat higher in private schools, but lower in Mali and Mauritius. In both Jordan and Morocco there are marked differences in all three domains and grades—private-school children outperform public-school children—while the opposite trend is found in all domains in Mauritius, where public-school children outperform private- (aided) school children. It is important to note that in both Mali and China the differences between private- and public-school children in the BLC domains are not as marked as in the other three countries.

The performance indicators chosen here to ascertain to what degree, in these five countries, the targets set at Jomtien are being met in the field of BLCs are just a first step for further in-depth analyses of the information-base developed through this joint project, and in particular, for the analyses of key BLC determinants.

Other findings. Due to the absence at this stage of reporting of a complete set of data from all four countries, we can only add to this progress report some findings on repetition and student age in Jordan and Morocco. Psychologists have demonstrated the effect on self-image of retaining students, and educational research has shown a high correlation between retention and later dropping out of school. A number of controversial issues thus arise, including whether it is beneficial to students and cost-effective to society to retain students.

In Jordan, students who had never repeated a grade performed significantly better than students who had repeated at least one grade between Grades 1 and 4. Age in grade is closely correlated with repetition rates in countries such as Jordan where most children enter formal schooling at the same age. There is strong evidence that older students, particularly those two or more years older than their classmates, internalize a wide range of negative feelings and are considerably more likely to drop out. In Morocco, the younger pupils (under ten years old) in Grade 4 had higher scores than 10- and 11-year-olds, suggesting that once children are behind they tend to remain behind.

Meeting the challenges

Even before all the results have been analysed, it is clear that the open-ended and flexible monitoring project approach—based closely on a common core of concerns and reinforced by frequent consultations with the central team at UNESCO headquarters as well as between participating countries and the two agencies (UNESCO/UNICEF)—has helped to avoid many of the problems encountered in the traditional approach to educational assessment.

The WCEFA provided a global framework for the development of education by the year 2000 and beyond. Monitoring EFA goals is seen as an important WCEFA follow-up mechanism. Although it is imperative to monitor the quantitative expansion of education and the related financial and personal inputs, considerable emphasis should be given to the effective monitoring of learning achievement.

Within the framework of the monitoring project, where monitoring EFA goals is approached from the standpoint of student learning achievement, the first five 'pioneer' countries have designed their projects, developed indicators, identified sampling populations, pretested and administered their tests and questionnaires, and begun to act on the results obtained. But the real 'results' of the project—national capacity-building, enhanced further by the multiplier effect—are already evident, in every step of the project, as described in this chapter.

If the ultimate objective of any follow-up activity to the WCEFA is to improve the quality of education-for-all, then the monitoring project's overriding aim of building national capacities is a crucial step along the way. Excellent interagency cooperation between UNESCO and UNICEF, together with the TCDC components emphasized at the initial stage of the project have laid the groundwork for the smooth integration of the next group of countries participating in the project. Already, we have seen the successful experiences of China being used to launch the project in Sri Lanka, and the national capacities developed in Jordan being used to train Omanis in the design and implementation of their project.

But most important of all, the foundations have been established for other countries to set up similar monitoring activities, based on the experiences of these first countries. A handbook containing the findings, methodology, and instruments of the project, along with insights into the lessons learned by each of the five pilot countries during the first phase, is in press (Chinapah, 1995). Through its emphasis on capacity-building, promoting TCDC, and encouraging a country-specific, feasible and sustainable approach to monitoring, the groundwork laid during the three-year project represents a major contribution towards meeting the challenges of the future.

References

Biazen, A. and Junge, B. (1988). *Problems in Primary School Participation and Performance in Bahir Dar Awraja*. Addis Ababa: UNICEF/Ministry of Education.

Carron, G. (1992). *Analysing Inequalities in Education: Some Comments on Indicators*. Paper presented at the International Consultative Forum on Education for All. Paris: UNESCO.

Chinapah, V. (1983). Performance and participation in primary schooling. *Studies in Comparative and International Education*, No. 8., University of Stockholm.

Chinapah, V. (1995). *Monitoring Learning Achievement: Towards Capacity Building—A to Z Handbook*. Paris: UNESCO.

Elley, W. B. (1992). *How in the World do Students Read?* The Hague: IEA.

Eshiwani, G. (1983). *A Study of Women's Access to Higher Education in Kenya with Special Reference to Mathematics and Science Education*. Nairobi: Kenyatta University College.

Postlethwaite, T. N. and Ross, K. N. (1992). *Effective Schools in Reading: Implications for Educational Planners*. The Hague: IEA.

UNESCO (1992). *Approved Programme and Budget for 1992–1993*. Paris: UNESCO.

UNESCO (1993). *Report on the International Workshop on Survey Methodology*. Paris: UNESCO.

UNESCO (1994). *Monitoring Education-For-All Goals: Focusing on Learning Achievement. Progress Report on the Project's First Five Countries: China, Jordan, Mali, Mauritius and Morocco*. Paris: UNESCO, Division of Basic Education.

CHAPTER FOUR

International comparisons of student achievement

Harvey Goldstein

Introduction

This chapter discusses some of the key technical procedures which have underpinned international comparisons of educational achievement; namely those concerned with sampling and population definition, translation, scaling and statistical modelling. The chapter does not intend to provide a detailed summary of the findings from comparative studies. It is concerned with the ways in which any such findings can be interpreted, and will draw lessons from existing studies in order to make recommendations for the future.

It is clear that there are political constraints on international comparative studies which are a source both of strength and weakness. They are useful insofar as governmental funding and support for these studies tends to ensure ready access to educational institutions and policy discussions; they are a drawback when they dictate a narrow view about which comparisons are important, and how findings should be presented. I do not deal directly with such political issues, but it should be appreciated that many of my conclusions will have political as well as scientific implications for the conduct of these studies.

The first three parts present a brief history of international comparisons, the organizations involved, and a summary of the measurements which have been made.

Historical background

There have been a number of small scale, limited and usually informally structured comparisons of achievement among countries. There are only a few studies which merit serious attention, however, namely those carried out under the auspices of the IEA and by Educational Testing Service (ETS). These are summarized in Table 4.1.

It is apparent from this table that there has been an increasing country participation rate from the first study in 1964 with an increasing frequency of studies in the late 1980s and early 1990s. It is also clear that the most common curriculum areas covered are science and mathematics. The most popular ages surveyed are nine to ten years and 13–14 years. Additional information gathered in the various studies includes curriculum, organization, teacher experience and some characteristics of students and schools. With some exceptions, such as the IEA second mathematics study, there is little reliable background extra-institutional information about the characteristics of a student's parents, home amenities, etc., and this limits the kinds of causal explanations which can be offered.

Table 4.1   Major international comparative studies of educational achievement

Years of data collection	Sponsor	Ages of pupils	Curriculum topics (number of countries)
1960	IEA	13	Mathematics, science, reading comprehension, geography, non-verbal reasoning
1964	IEA	13, FS	Mathematics (13)
1970–72	IEA	10, 14, FS	Science (19), reading comprehension (15), literature (10), French and English foreign languages (18), civic education (10)
1980–82	IEA	10–14	Classroom environment (mathematics, science, history) (10)
1982–83	IEA	13, FS	Mathematics (20)
1984	IEA	10, 14, FS	Science (24)
1984–85	IEA	10, 14–16, FS	Written composition (14)
1988	IAEP	13	Mathematics (6), science (6)
1988–92	IEA	10, 13	Computers in education (23)
1988–95	IEA	4	Preprimary education (14)
1991	IEA	9, 14	Reading literacy (31)
1990–91	IAEP	9, 13	Mathematics (20), science (20)
1993–98	IEA	9, 13, FS	Mathematics, science (40–50)
1995–	IEA		Second language

FS denotes the final year of secondary education, differing among countries.

IEA is the International Association for the Evaluation of Educational Achievement.

IAEP is the International Assessment of Educational Progress.

Organizations

The organization which dominates these international comparative studies is the IEA. The IEA consists of a set of member institutions (upwards of 50), usually one from each educational system which send representatives to an annual General Assembly which is the decision-making body in which every system has a single vote. (In some nations, such as Belgium and Canada, two separate educational systems are operated along cultural lines.) The IEA has a small permanent secretariat, based in the Hague, Netherlands and an elected chairperson and standing (executive) committee. In addition it has a technical advisory committee drawn more widely than the member institutions. For each project an International Steering Committee is set up together with an International Coordinating Centre whose head is formally executive director of the project, and is responsible for ensuring that the data collection instruments are prepared and the data processed into a form suitable for international and national analyses. In each participating country there is a National Project Coordinator who usually belongs to that country's member institution.

Each member institution of IEA is expected to have links with the country's policy makers as well as having research expertise and access to schools. The funding to enable a country to participate in a study almost always comes from central government sources. It is principally in this way that governments make their interests known. Very often, however, the costs of funding pilot or feasibility projects is found from private foundations or particular governments, typically from Western Europe, Australia, the United States or Japan. A description of the IEA structure and studies is given by Hayes (1991). The importance of government funding means that some of the poorest countries are unable to participate because government resources are too limited.

The IAEP has been organized by ETS, the major testing service in the United States, building upon that organization's experience of managing the US National Assessment of Educational Progress (NAEP). It has carried out just two major surveys, almost entirely concerned with science and mathematics. Unlike the IEA which is fundamentally a democratically structured organization, the principal decision-making functions of the International Assessment of Educational Progress (IAEP) are located within ETS which also assumes responsibility for the major analyses, although some participating countries carry out their own. It appears (1995) that there are no plans for future studies.

Instrument development processes

The process of developing instruments for the collection of achievement data and other characteristics of students, teachers and schools is clearly of the greatest importance. I shall describe in turn the various types of instruments and the manner of their development. This part will mainly draw on the IEA experience since this is far more extensive and better documented than the IAEP. Further details of some of the technical procedures can be found in Keeves (1992a).

Student and parental characteristics

Much of the information about students is obtained from questionnaires which, for example, ask about the amount of homework undertaken, or attitudes of parents to the student's studying. There are standard procedures for the general quality control of questionnaires as follows.

First, it is necessary to formulate objectives and then to start the process of translating these into questions or scale items. In an international study this will involve representatives of several countries in the pilot stage, with approval and modifications being sought as to the general applicability of the tasks to be placed before students. Fundamental to this process is that of language translation, discussed in a following section. Nevertheless, even though questions may be correctly translated, their interpretation may differ. Even apparently 'hard' data such as the number of older children in a student's family will depend on the interpretation of 'family', for example whether it includes older 'adult' children not living with the student or children of a previous parental partnership. Such problems will tend to assume greater significance in a self-completion questionnaire to students than in the more common interview survey. Often, it will be at the analysis stage that such problems are revealed. It would assist interpretations of findings if these problems were documented so that an attempt could be made to understand the problems of comparability in questionnaire information.

Although the emphasis in international studies is on comparability of information so that a common interpretation can be made, there are some variables where this may not be possible, nor desirable. Perhaps the most important case is that of social status, where a common measure applicable to all kinds of economic system and culture will be unsatisfactory so that each country will be obliged to form their own most appropriate one— although there is a great deal of debate within many countries about how this might be done. The important issue here, in terms of data analysis, is the extent to which an appropriate measure of social status is associated with achievement. Comparisons between countries might then be made in terms of the relative strength of such relationships.

Where international surveys are part of a sequence, the need for maintaining comparability of questions over time is important, but there are particular difficulties when new countries join a sequence of surveys so that modifications may become necessary.

The emphasis on comparability across countries and time is one that seems to underlie almost all the activity of international surveys, and is derived largely from the early desire to make unambiguous comparisons. I shall be returning to this point later in the discussion of interpretation, but for now simply note that in the design of questionnaires, strict comparability often may be unattainable, but that this does not rule out the possibility of useful analyses.

School, classroom and curriculum measures

The IEA classroom environment study was unique in attempting to measure classroom processes directly by observing and coding student and teacher activities over time. Such information can provide very valuable contextual data for comparative analyses, but requires careful training of observers and is relatively expensive. More usually, information about the schools, teachers and curriculum policies is obtained through questionnaires. In addition, there will be information about school type, size, etc. and in some systems with prescribed curricula the 'intended' curriculum structure.

The IEA has always shown interest in descriptions of the curriculum in different countries both for their intrinsic interest and to help in contextualizing achievement results (see for example, Travers and Westbury, 1989; Finegold and Mackeracher, 1986). The US government is funding a survey of mathematical and science opportunity (SMSO) to attempt to develop data collection strategies for obtaining 'opportunity to learn' (OTL) data, that is information on the extent to which the IEA assessment item topics have been covered in classrooms.

In evaluating the performance of any group of students, it is important to know what their curriculum exposure has been, and this will be discussed in a later section. In addition, information about the extent to which different groups of students experience different topics and how this relates to overall curriculum goals is of great interest. The current concern of the IEA with this issue, therefore, is welcome. Nevertheless, there are several outstanding problems to be tackled. For example, the reliability of information from questionnaires is not only likely to be low, but also to vary from education system to system. What works reasonably well in systems with clearly described curricula may work badly in systems with more decentralized and informal curricula. More seriously, OTL data is normally collected on a class or group basis, whereas there may be variation in exposure from student to student within a group.

In the important area of curriculum description there is interest in information on the intended curriculum, namely that which is described in official, local or central documents. The implemented curriculum is that which is provided in schools by teachers and texts which interpret or modify the intended curriculum. The relationship between these two is in general not well understood and OTL data in the IEA surveys provide an important opportunity for investigating it. Finally, there is the attained curriculum, which, loosely, can be described as that which has been absorbed by students. It is this which the assessments themselves are attempting to measure.

In approaching the comparisons of curricula, there are considerable complexities, since each educational system embodies its own cultural assumptions which interact with documentary descriptions and classroom practice. To understand curriculum differences it is also necessary to understand the cultural contexts. Leung (1992), in a detailed comparison of the mathematics curricula in China, Hong Kong and England, describes in detail how cultural assumptions, transmitted via teachers and others, can affect the implementation of a curriculum.

From teachers, information can be obtained about qualifications, experience and attitudes. From schools, information can be obtained about organization, student grouping, resources available, relationships with parents, staffing, etc. Naturally, much of the information required about the class and school contexts is difficult to measure precisely using questionnaires, especially since much of the information is retrospective. Furthermore, as with curriculum information, this information needs to be contextualized within cultural settings.

Sampling procedures and population definitions

From Table 4.1 it is clear that there are certain favoured ages. There appears to be a lack of studies in the early years of schooling, between the ages of five and eight, and likewise for ages 11 and 12 which correspond to institutional transition ages in many countries. In fact, the populations are defined in terms either of school year or grade. This creates difficulties for comparisons, since school and country policies vary with respect to grade or year promotions; in some countries the whole year group moves together, whereas in others some students repeat years or grades. This problem will be discussed in more detail later.

Most of the existing studies have concentrated upon cross-sectional comparisons, that is studying students at a single time. There is a deficit of long-term longitudinal studies which could shed important light on the factors associated with student progress and possible causal mechanisms. The IEA, for example in the second mathematics study, followed up a subsample of students over a nine-month period in a small number of countries, but such short-term studies are of limited value when the curriculum, and programmes of study in general, are designed to cover longer time periods.

Within individual countries (see Mednick and Baert, 1981) successful longitudinal studies have been carried out and there would seem to be good reasons for the IEA to attempt the same, although such studies do require large amounts of resources. While successive surveys of literacy or science may provide interesting snapshots, the scope for making causal inferences is severely limited. Longitudinal studies can begin to answer questions about student mobility and its causes, changing performance differences between groups as they progress through the system, and many other issues of considerable significance for educational policy and theory.

The sampling procedures adopted by IEA and IAEP involve, on the whole, standard applications of sample survey methodology. The primary sampling unit usually is the school, and schools typically are stratified, for example by type, region and size. In international studies it is often difficult to ensure uniformity of sample design across countries, so that different weighting procedures may be necessary prior to comparative analyses.

Particular problems can arise when sampling students in the final year of secondary education. If the population of interest is those students in school (or other educational institutions) then there are no novel problems. For some purposes, however, the population definition will be wider, for example to include young people in training activities, and even a whole cohort of a particular age. The sampling then has to encompass these groups outside institutions and becomes more difficult and expensive. One solution to this problem is to define an age cohort of interest and to identify the sample individuals while they are still at institutions, that is prior to the compulsory leaving age. Such a sample would then be followed-up to the age of interest. In addition to the sampling issue, one advantage of such a procedure is that longitudinal information becomes available which might be expected to be of considerable interest. The principal disadvantage is that it requires a longer time span. Such a scheme does not seem to have been adopted for international comparisons, but seems well worth exploring.

Response rates

The response rates, for both schools and students, vary from country to country and age to age. For example, in the second IAEP mathematics study, at age nine the overall student response rate varied from 53% in England to 99% in Taiwan and at 13 years from 47% in England to 98% in Taiwan. In the second IEA science study (SSIS) the student response rate for ten-year-olds varied from 54% in Norway to 99% in Japan and Korea, and for 14-year-olds from 53% in England to 100% in Korea. Because lower response rates are generally associated with increased bias, it is important that any comparisons between countries with such different response rates are treated cautiously. This is underlined in the IAEP surveys where countries with low response rates or with restricted sampling frames such as those covering largely urban areas, are listed separately. The IEA summary science report, however, (Keeves, 1992c) pays scant regard to this issue, although it mentions that there are comparability problems 'which make it difficult to compare the performance of students'. For example, in comparisons between the first and second IEA science studies, the response rate among ten-year-olds rose from 49% to 84% in Italy, and the response rate for 14-year-olds in Sweden dropped from 91% to 50%. In spite of this the report goes on to make comparisons without attempting to allow for these problems.

The problem of non-response in surveys is a difficult one, and although it is recognized by those designing comparative studies, it is still an area where such studies are weak. In some cases it may be possible to measure certain characteristics of the non-responding schools and students and to use these as a check on the obtained sample. The use of call-backs with requests for basic information from schools is worthwhile, as is the use of nationally available information on pupil–teacher ratios, teacher qualifications, examination results, etc. Alongside other estimates of statistical uncertainty (see below) comparative tables should provide estimates of possible non-response biases.

Age and grade sampling

The IAEP samples were defined by the year of birth of the students. Thus, for example, for the 1990–91 survey the target population was all children born in 1977 and for most countries these were measured in March 1991. This yielded an age range of one year with a similar distribution of ages within countries. For a small number of analyses, results have been reported separately for the two principal grades into which the students fell. In the second IEA science study the actual mean ages of the country samples for the 14-year-old students ranged from 13.9 years to 15.1 years (with an outlier at 16.0 years). In this study, most of the country samples were from Grade 8 or 9, and in some cases the mean age of some Grade 8 students was higher than the mean age of some Grade 9 students.

Both length of time in school, which is what grade level is intended to measure, and age itself will influence achievement, attitudes, etc. Furthermore, different systems have different policies about whether weak students should repeat grades. Comparative analyses need to take careful account of these problems, ideally adjusting results for age, grade level and the extent to which students have experienced grade repetition or promotion. Information on grade allocation policies is of interest as a possible explanation for country differences, and this also raises a number of interesting issues about 'compositional' effects, namely how individual achievement is affected by the characteristics of the other students in the same class. This in turn requires the use of 'multilevel' statistical modelling which is discussed below.

To date, little attempt seems to have been made to develop reliable procedures for simultaneous age and grade standardization of information prior to reporting, and most published reports, unfortunately, do not appear to regard it as a serious problem. There are, of course, difficulties in making proper adjustments for both age and grade level (McDonald, 1992) and this is an important area for further research. A recent report on the reading literacy study (Elley, 1992, Appendix E) did carry out some limited age adjustments and showed how this affected some of the country comparisons. Unfortunately, age adjustments were not used for the comparisons in the body of the report. Fortunately, the third international mathematics and science study has proposed that students in two adjacent grades are sampled, so allowing exploration of combined age and grade effects.

Translation procedures

In the IEA as well as the IAEP, the first versions of all instruments are usually in English, although not necessarily devised solely by native English speakers. The problems of translation of questionnaires and tests have been studied and discussed by a number of researchers, some in the context of the IEA studies (see for example, Brislin, 1970; Little, 1978; Purves, 1992). A number of guidelines have been evolved, along the following lines.

A basic requirement when translating from a source to a target language is to back-translate the text from the target to source language and to compare the original with the back-translated version. Bilinguals are commonly used to doing this and in some cases complex experimental designs with several translators have been utilized to study the effects of factors such as textual content, translators' experience and familiarity with subject matter. It seems clear that all these factors can influence the quality of a translation. Moreover, even where there is a good match between the original and back-translated version, the target version will not necessarily be an appropriate translation. This might occur, for example, because a single source word can have several translations in the target language, each of which would be back-translated into the original source word, yet each target language word can nevertheless have a somewhat different meaning.

An interesting example occurs with Japanese which has context-specific number systems. A number from an English-designed test could have various target translations depending on the context yet all be translated back into the same English number. Another example is given by Little (1978) and concerns the use of the word 'expect' when asking students about their future careers. In English there is a difference between 'expect' in the sense of 'wish to' and 'expect' in the sense of 'predict'; that is between hopes and predictions. Some other languages do not distinguish these meanings, partly it seems because the social and cultural conditions make such a distinction unnecessary when ambitions are strongly determined by practical realities. In both these examples, although the goal of exact translatability is unreachable the language differences lead to substantively interesting questions, and I shall return to this point shortly. It is worth mentioning, however, that where unique one-to-one equivalences between all translated words or phrases is required this may be achievable only at the expense of eliminating useful test items or questions.

It is generally agreed that passive constructions, pronouns and complex structures should be avoided. There seems to be little research, however, on the effect of such an injunction upon the meaning in languages other than English. For example, simple structures in English do not necessarily carry over into simple structures in other languages, especially pictographic languages such as Chinese. Translators are also well aware that within countries, there are usually dialects, some of which may be unfamiliar, yet important if representative population groups are to be sampled. It is also difficult to equate levels of concreteness and abstraction in two different languages. McLean (personal communication) quotes the example of a French translation of a mathematics test which satisfied strict quality controls, but was unable to deal with items which were judged to be more abstract than their English counterparts. Hanna (1993) describes a study using six bilingual French–English educators who performed a content analysis of 174 items in the IEA second international mathematics study, all of which had been back-translated. They reported that 70 of the items were found to differ in significant aspects in the two languages. The examples Hanna quotes suggest that many of the differences are potentially avoidable, but to do this would require considerable resources to implement on a large scale.

In the practical situation of a study, operating under constraints of time and resources, it is difficult to take account of all the problems associated with translations. In some cases (Rosier, 1987) lack of resources has prevented some countries even producing back-translations. Among other things, such practical issues imply that the analysis of comparative studies needs to be sensitive to potential translation biases. This will be especially so where large or unexpected differences occur, and translation problems need to be eliminated as explanations. For example, it is conceivable that the context-specific nature of a Japanese translation of numerical information may contain information which facilitates a correct response. This strengthens the case for complete documentation of all the study materials, administration instructions etc., and public access to these.

The position of English as the source language for most comparative studies raises some special issues. One of these, the assumption about simple structures, has already been mentioned. There are also other concerns. Not only is English the main source language for the instruments, it is also the common language of discourse among those jointly designing, discussing and analysing the studies. In the IAEP, the fact that it was organized by ETS implies an inevitable dominance of the concerns and cultural values of particular groups in one country. Yet even in the IEA, with its more democratically multinational structures, the requirement for country representatives to have a working knowledge of English in order to take part in joint discussions necessarily implies a similar, if not so pronounced bias.

It is, of course, difficult to quantify the extent of such biases. The English-speaking psychometric tradition is so universally dominant, that its assumption as a starting point for discussions about educational measurement is usually simply taken for granted. Nevertheless, it is possible to carry out research which would throw light on this issue. Languages other than English could be chosen as starting points for test and questionnaire development and the resulting instruments used alongside the English-originated ones.

Psychometric approaches to translation

Recently, there have been suggestions that there are psychometric 'solutions' to judging the effectiveness of translations (Hulin, 1987; Hambleton, 1992). In essence these authors propose the following basic psychometric model.

The test item patterns from random samples in the source and target populations are compared to see if they are similar. For example, if the items are ranked in order of difficulty based upon the proportion of students answering them correctly, then one criterion would be based upon discrepancies in the rank orderings. Similarly, if the (biserial) correlation between an item response and total score differed in the two populations this would be viewed as evidence for possible translation problems. Variations upon such criteria are often used, for example based upon non-linear weighted functions of item responses, but the principle is the same.

This technique is akin to procedures which have been suggested for detecting 'biased' items when comparing subpopulations, for example defined by gender or ethnicity. The difficulty is that there is no way of knowing whether a few 'aberrant' items present translation or other problems or whether they are in fact valid achievement indicators measuring real population differences. The usual psychometric procedure for resolving this dilemma is to make the assumption that the set of item responses can be modelled in terms of a single one-dimensional student 'ability' or 'trait', in a sense which is discussed more fully on p. 73. Such an assumption, unfortunately, merely restates the dilemma, this time in terms of whether an aberrant item should be regarded as problematic or whether the item set is legitimately viewed as spanning at least two dimensions. This is not to say that such analyses cannot be used to provide suggestions about interesting population differences, but rather that they cannot properly be described as tests for translational validity. A more detailed discussion of such psychometric tautologies is given by Goldstein and Wood (1989).

Finally, it does seem reasonable to ask whether in all cases perfect or near perfect translation is worth aiming for. The inherent variation in language structures in some cases seems to preclude this anyway, and in other cases the practical difficulties deny full knowledge of whether the goal has been achieved. Instead, we should perhaps regard the translation issue as belonging, at least partly, to the stage of data interpretation. The goal of trying to render tests and questions equivalent is a sensible one, so long as it is recognized that subsequent analysis may provide further insights and understandings about linguistic and cultural differences.

Data processing technicalities

Since the early days of international studies, the computer revolution has transformed the data processing and analysis of large scale surveys. Data transfer from test booklets and questionnaires and other instruments can be carried out rapidly, and the process of cleaning data prior to analysis likewise has been speeded up. This has been demonstrated in the reports of the IAEP first and second mathematics and science surveys (Lapointe *et al.*, 1989, 1992a, b) where initial analyses were published little over one year from the start of the survey. These studies utilized computers at most stages of piloting, administration and analysis within participating countries as well as centrally.

In the past the IEA publication time scale has been longer. For example the second international mathematics study began to produce fully comprehensive country comparisons some three years after data collection started (Robitaille and Taylor, 1986). More recently, however, the IEA computers in education study (Pelgrum and Plomp, 1991) produced a comprehensive report within two years of starting to collect data which included some quite complex statistical modelling. Likewise, the IEA reading literacy study collected data in 1991 and produced a first summary report in mid-1992. It seems not unreasonable to expect future surveys to produce useful summary reports not more than a year after data collection ends and to make data available for secondary analysis shortly afterwards.

Data scaling and data interpretation

A prevailing assumption behind all international comparative studies has been that they exist principally, if not entirely, in order to describe country differences. The desire to explain differences, for example in terms of curriculum exposure, teacher attitudes or cultural expectations has always been of concern and a relatively recent development has been the use of powerful statistical modelling for this purpose (see for example Pelgrum and Plomp, 1991 and Keeves, 1992c). Without an attempt to provide such explanation the descriptive statistics have little real use, other than as political propaganda. This is perhaps most evident in the analyses produced by the IAEP for science and mathematics achievement (Lapointe *et al.*, 1989, 1992a,b).

The first international report on the IAEP assessment of science and mathematics in 1988, was based upon the US NAEP which was carried out under the auspices of ETS. In a slim but well presented and speedily published booklet, ETS presented comparative information about the average performances of each of five countries under various topic headings. For example, in mathematics the percentage of items correct for each country is reported for the topics of 'number', 'relations', 'geometry', 'measurement', 'data organization', and 'problem solving'. In addition the average total number of items correct is reported. As well as this there are tables comparing the reported frequency of classroom mathematics and science activities, amounts of homework and attitudes of students towards mathematics and science. There are also some comparisons based upon

OTL, that is an average measure of the students' exposure to the topics being tested.

Except for a couple of instances, there is no attempt to interrelate factors. For example, it is extremely difficult to establish the fact that there is an association between OTL and performance on each topic (Wolfe, 1989). While the report does present results for separate topics, its main emphasis is on the overall science and mathematics 'proficiencies'. These are simply (weighted) averages of the subtopic scores with the weights approximately reflecting the number of items in each subtopic scale. Thus, in mathematics, since there are 24 number items out of the total of 62 and only eight problem-solving items, the proficiency scale is much more heavily weighted toward the former. The report itself fails to comment on the implications of this. Rather it seeks an interpretation of the proficiency scale by adding verbal descriptions to it, corresponding to particular scores based upon the observed performances of individuals achieving those scores. Thus, a score of 300 is said to correspond to students (at Grade 8) who 'can add two-digit numbers without regrouping and solve simple number sentences involving these operations'. The report claims that these descriptions can inform the reader about what children at that score point 'know or can do'. Despite a caveat in the introductory section, there is little in this same report which tries to convey the tentative nature of international comparisons, and the problems of translation and interpretation which are well recognized by those responsible for designing and analysing the assessments. I now turn to some of the interpretation issues raised by this IAEP report, bringing in IEA material also.

Opportunity to learn (OTL)

Comparing educational performance among population groups is a somewhat pointless exercise unless it can be contextualized by measuring the exposure students have had to relevant learning experiences. Clearly there are many influences on performance, but if education has any effect the exposure to a topic should be associated with performance on that topic. Thus, the information that the United Kingdom does well in problem solving should be read in conjunction with the relatively high exposure that UK children receive. Indeed, such exposure information may be of more use for many purposes than the performance data. In fact, from IEA surveys, although OTL is associated with achievement, the relationship does not always appear to be very strong (Goldstein, 1987: Chapter 5). Moreover, the relationship seems to vary across countries.

The principal difficulty with existing measures of OTL is that they tend to be rather coarse, measured for a group of students rather than each one individually, and based upon retrospective data, namely the responses of teachers. It is to be expected that these circumstances will underestimate markedly any relationships which exist. In view of the importance of measuring OTL, one would hope that future resources will be directed at obtaining reliable individual student-level data, and the SMSO study, mentioned earlier (p. 62), promises to be a useful starting point.

Aggregated scales

One of the more misleading presentations of results of comparative studies is the emphasis given to aggregate scores of 'mathematics' or 'science'. Such scale scores typically are formed by averaging the responses for all the items in a subject area. This has two principal drawbacks. The first is that much of the real interest lies in individual topic areas and the second is that this reflects the weightings of topic items chosen by the test constructors. It has already been pointed out how in the IAEP first study, the implicit definition of 'mathematics' was weighted by number items, and this likewise has been a persistent problem in the reporting of IEA results.

The choice of items to be used in assessing, say mathematics, is the result of a negotiation among the participants in a study. The 'core' set of 'consensus' items agreed upon as common to all countries are those upon which international comparisons will be made. Yet because these represent a compromise, they may not be representative of any single country's overall intended or implemented curriculum. Nevertheless, they will be more representative for some countries than others. Thus, for example, in the IAEP first mathematics survey already discussed, those countries where the curriculum emphasizes numerical competencies as distinct from problem solving will be relatively advantaged in comparisons of overall mathematics scores. As Wolfe (1989) points out, if different weighting systems are used for the components of mathematics, the relative position of countries will change, and he quotes the example of England and Wales which move up the country rank order if an equal weighting is applied to topic areas. Westbury (1992) further discusses this issue and compares student achievement in terms of the curriculum coverage of Second International Mathematics study items in Japan and the United States. Unfortunately, his conclusions need to be treated with caution since his analysis does not properly adjust for topic selection factors at the student level, and also confines itself to adjusting for pre-existing achievement at the class rather than student level.

It seems clear that the very notion of reporting comparisons in terms of a single scale, for example of 'mathematics' or 'science', is misleading. Purves (1992) makes this point strongly with respect to writing proficiency, where he suggests that at least three separate dimensions are present and that student responses have to be interpreted in the light of cultural differences and expectations. He also emphasizes the subjective nature of choice of items in any test, and his reservations about interpretations can be made for the other subject areas. Likewise, Swain (1990) points to the contextual influences of item characteristics on student responses and the complex multidimensional structures involved in second language testing.

It appears to be somewhat pointless to devise separate test forms for components of science or maths or language if reporting is then undertaken principally in terms of a single scale. One possible alternative is to report several scales, each using a different weighting, but while this seems worth investigating, it may be somewhat confusing for most readers. What then is the appropriate level at which results should be reported? At one extreme it is possible to report on each assessment item separately. This has certain merits, and there is a strong case for item level analyses to be available. Yet, again, typically there are natural groupings of items covering specified aspects of the curricula which can form meaningful reporting levels. If this is to be done then it is also important that readers of reports have easy access to all the constituent items, in the relevant translations, and not merely a sample set.

If the analysis of subscales of achievement is to be pursued, then it will be fruitful also to study the interrelationships between scales. That is, the extent to which performance on say, problem solving in mathematics, is correlated with data analysis proficiency, and whether these relationships differ from country to country.

Statistical scaling

Despite the argument of the previous part in favour of disaggregated reporting, a number of data analysts claim to have developed single scales for 'science' or 'mathematics' or 'language' which would allow valid comparisons in terms of a single scale value, irrespective of which subset of items from a larger collection was used in the assessment or what relative weightings were used for different components. Thus, the first IAEP science and mathematics surveys use of so-called 'item response' scaling to produce single proficiency scales in mathematics and science, and Keeves (1992a, b) argues for such scales and presents one such scale for science achievement in the IEA first and second science studies. Among other things, it is claimed that such scales allow comparisons of national achievement over time, independent of curriculum or cultural changes. In essence the argument is as follows.

In order to illustrate the procedures, the scale developed for the IEA science studies will be used. The assumption is first made that all the items under consideration are reflecting a single underlying 'trait' or 'dimension'. The general procedure is to ignore *prima facie* evidence for separate scales but rather to see whether, after constructing the scale, the data themselves provide evidence for rejecting a single scale. In the IEA first and second science studies, the scale was developed from the 14-year-olds science achievement test items which were common to both surveys (Keeves, 1992c).

For each of these common items the basic assumption is made that any change in the proportion of correct responses over time is a reflection of changes in the population rather than, in an alternative sense, changes in the facility of the item. Thus, if the correct response rate for a physics item significantly increased from 40 to 50% between the first and second science surveys this would be interpreted as an increase in student achievement in this area of the physics science curriculum. At this preliminary stage, some items may appear 'anomalous', for example, for which the population response remains unchanged rather than increases as for the remaining items. A common procedure would be to eliminate such items as 'non-fitting' so that the scaling is then carried out on the remainder.

It is unnecessary to go into the details of the procedures by which final scale scores are produced, typically using time-consuming statistical modelling. In essence, however, for each survey one can think of making an estimate of the underlying trait of 'science' by calculating the average item score for each student—which is the average proportion correct if the items are simple pass/fail ones. A slightly more refined method uses a weighted mean where the weights are determined by the intercorrelations of the items. For one of the surveys, say the second science one, these student scores are then simply scaled so that they have a designated mean value (500, say) and a spread (0–1000, say). Once the equivalence between such a 'convenience' scale and the 'raw' student scores has been established, all the scores can be given a scale value.

Having established this scale, it can be extended to include new items, so long as they are assumed to belong to the same 'trait'. This is done by comparing student responses on the new items to student responses on the existing scale items so that each new item can be assigned a 'difficulty' value (and if the more refined method is used other characteristics such as 'discrimination') alongside the difficulties of the existing items. With this information the new, more extensive instrument can be used to assign scale values to students. The 'linking' of tests in this way can be carried on for more stages, and in the IEA science study the 10-year-old and 14-year-old tests for each survey were finally linked into a common scale. The results of such a procedure will sometimes be incorporated into a calibrated or scaled 'item' bank. From such a bank subsets of items can then be selected to form tests whose overall difficulties and other scale properties are regarded as known.

A technical description and evaluation of these scale creation procedures is given by Goldstein and Wood (1989). Before going on to look at how these scales have been used to interpret achievement, their limitations need be discussed. It should also be pointed out that the IEA science researchers are not alone in preferring such scales: they have been used extensively by ETS in the IAEP and the US national achievement survey, the NAEP. It has also been proposed that such scales are used in the third international mathematics and science study (TIMSS, TAC, 1993).

Limitations of item response scaling

A crucial assumption used in item response scaling is that of unidimensionality, defined as follows. If, for a set of test items or questions, the responses of a group of students are determined by a single 'trait' value, then that set of items is said to be unidimensional. In other words, the responses reflect the operation of one and only one underlying factor, be it 'reading ability', 'abstract reasoning' or whatever.

First of all, it should be noted that such a definition has to be population dependent. Thus a test may be approximately unidimensional in one group of students but clearly not in another. This is especially relevant in international studies where very different systems and cultures are operating. Because a set of relationships holds in one or more countries this cannot guarantee that it will do so elsewhere. In practice, of course, no set of items is perfectly one-dimensional, so that some statistical procedure has to be used for deciding whether a set of items 'approximates' unidimensionality, and thus involves subjective judgements about what constitutes an adequate approximation. In order to achieve a scale that approximates 'unidimensionality' those items representing 'minority' dimensions will have to be removed or suitably modified until they conform. This of course will tend to increase the unidimensionality of a test, but not necessarily its 'validity' or fitness for purpose, that is, its capacity to measure what is intended. This is seen easily in the following simplified example.

Suppose we have two sets of truly unidimensional items representing respectively dimension A and dimension B. A test constructor chooses a 50 item test using 40 items from A and ten items from B. She then carries out the standard 'item analysis' or 'item response theory' (IRT) procedures and discovers that the ten B items seem discrepant, that is they do not exhibit the behaviour of the majority. In accordance with common practice, and in order to obtain a unidimensional test these B items are omitted. A unidimensional test is obtained. But of course, it merely reflects the decision taken by the test constructor originally to weight the test with mainly A items. A unidimensional test would also have been obtained if the roles of items A and B had been reversed. In that case however the test would represent something quite different, for example ranking students differently and altering comparisons between population groups. This example is a simple one, but Goldstein and Wood (1988) show how the same principle applies quite generally. Swain (personal communication) gives an example from language testing where proficiency in 'ability to communicate' and proficiency in 'grammatical accuracy' differ markedly between French immersion students in Canada and students studying English in China. A test which was reduced to items reflecting just one of these proficiencies would thus disproportionately favour one group over the other.

In addition to these fundamental difficulties, the statistical procedures themselves are far from satisfactory. General so called 'goodness of fit' tests provide weak evidence for confirming unidimensionality unless they are concerned with contrasting a unidimensional structure with a specific multidimensional structure. Thus, in the above example, if we had information to suggest that the ten B items belonged to a separate dimension then a powerful statistical test could be devised for this hypothesis. Usually, however, such information is unavailable and a wide variety of possible alternative structures will have to be allowed for in the statistical test procedure. As a result, such 'non-specific' tests will often fail to detect a real multidimensional structure.

There is a further problem with almost all attempts to produce unidimensional scales. This derives from the fact that the data samples used tend to come from very heterogeneous groups. This means that where there are high intercorrelations among items, some of this will be due to other factors such as family background, and especially curricula differences. There are hardly any studies which seriously have attempted to study this issue by 'partialling out' such factors before reaching conclusions about dimensionality. The IEA studies in fact have relatively good data for this and the international facet would make such analyses particularly valuable.

We see, therefore, that the assumption of unidimensionality should be handled with care, and that despite a high level of statistical sophistication, both the objective and subjective intentions of the test constructors remain paramount. Claims for having established unidimensionality should be treated cautiously.

In the light of the discussion of scale construction and as with the case of mathematics discussed earlier, any overall scale is best viewed as a particular weighted average of its separate components with no other special meaning. Furthermore, there may be a serious conflict between claims for a single unidimensional scale while at the same time reporting separate components. This is illustrated, for example, by Keeves (1992c) who presents results comparing countries on a single science scale as well as in terms of separate components such as 'reasoning' and 'investigation'. On the results for the separate components the countries involved have differing rank orders, which suggests that a single international scale is highly implausible.

Time trends

While most attempts to scale test items across time have been concerned with producing general unidimensional scales and so left themselves open to the criticisms outlined above, it would in principle be possible to confine such attempts to narrowly defined, and hence perhaps truly unidimensional traits. Unfortunately this too runs up against logical difficulties, as follows (see also Goldstein, 1983).

Returning to the item whose facility rises from 40 to 50% from the first to the second occasion, how do we interpret this? It has been pointed out that item response models interpret this as a shift in the population's propensity to achieve success on the item. Yet just as easily we might assume that the population had not changed in any way, but that the item had just become 'easier'. It is possible to imagine situations where this might occur, such as the recent incorporation into common language of words used in the test. The reverse situation can also occur where an item can become more difficult because, say, the school curriculum has changed. These considerations will tend to apply only over long time periods, and over short time periods it may be reasonable to suppose that such factors are relatively unimportant. The problem, however, is that it is the longer time periods which are usually of most interest. Even over short time periods, however, serious problems can arise which will be discussed below.

The point is that it is impossible to resolve the issue of whether, in some absolute sense, an item retains its characteristics and the population changes or vice versa, or perhaps a mixture of both. In some circumstances it may be possible to reach a measure of agreement about the interpretation of any changes, but there can be no purely *technical* solutions to this duality of interpretation. What can be said is that on a chosen set of test items—those that happen to be common on both occasions—achievement has changed in particular directions. Simple interpretations of such changes (Keeves 1992a, c) are therefore uninformative. Among other factors which need to be studied is that of the continuing relevance of each of the common items to each country's curriculum. In addition, we would need to understand why the particular common items were chosen by the test constructors and whether the mechanisms of choice could have led to items 'biased' in one particular direction.

It seems that this duality problem is not well understood. For example, Keeves (1992c: 265) points out that 'to rely on the items that are common to the two occasions for any comparisons made, must likewise be considered to lead to an incomplete and inadequate assessment of change'. Yet he also claims that item response models allow the construction of a scale that is 'valid across countries and over time'.

Finally, recent empirical investigations have thrown some light on another aspect of item response scale construction techniques, namely that of item parameter invariance. The assumption underlying most of the psychometric models for test item responses is that the characteristics of a test item, for example its difficulty or discrimination, are constant and uninfluenced by different contexts. Thus, the ordering of items in a uni-dimensional test will not change the item parameter values, nor will the incorporation of new items in a test. (Note that this is not the same as the assumption of item response independence which states that for a given individual and a given test, the probability of a 'correct' response to an item in the test is independent of responses to any other items).

This assumption of item parameter invariance is extremely important for international comparisons where tests are often augmented by locally introduced items, and most importantly where comparisons across time are attempted using a set of items common to tests at each occasion, but coexisting with different other items at each occasion.

The recent evidence which casts doubt upon this assumption is that from the US NAEP, administered by ETS. It was found, upon comparing the results from the 1984 and 1986 surveys on the basis of a set of common items, that there were dramatic falls in performance for nine-year-olds and 17-year-olds. Because this was regarded as extremely unlikely, an extensive investigation was held to study the reasons (Beaton and Zwick, 1990). The conclusions were that the students' performances on the common items changed according to the context in which the items were administered, that is, how and where they appeared in the test booklets. In brief, 'when measuring change, do not change the measure'. The report urges considerable caution when contemplating the measurement of change over time and is clear that no satisfactory procedure for so doing is available.

Statistical modelling

There have been some attempts by IEA to use elaborated statistical models to explore the data. The Use of Computers study (Pelgrum and Plomp, 1991) uses structural equation models to explore the structure of data related to the implementation of computer education and there are examples of path models and some use of OTL information. The majority of analyses, however, concentrate on country comparisons and simple group differences such as those between males and females. The IAEP analyses (for example, Lapointe *et al.*, 1992) present aggregate level comparisons between test scores and other variables such as hours spent on homework, and also present crude indications of the strengths of relationships within countries between students. These analyses look at no more than two factors at a time and so are extremely limited in terms of explaining country differences.

To date there has been very little attempt to fully model the within-country variation in test scores and other variables. While the analyses generally have been careful to take account of the complex sample designs, they have not attempted explicitly to study the way in which achievement, attitudes, etc. vary from school to school, or area to area.

It is particularly important to develop explanatory models which attempt to explain statistically observed relationships. Such relationships might be those between, say, OTL and achievement or between achievement and reported hours of homework. What is of real interest is to explore reasons for such associations in terms of other measured characteristics. These might be the experience of the teachers, curriculum variables or the home background of the students. It is also important to study whether explanations for these relationships differ among countries and then attempt to understand why.

Clearly much of this kind of analysis could be carried out by researchers not involved in the original studies. To make this feasible, however, requires easy accessibility not only to the data files, but also to the original test forms and questionnaires and implies a high level of data organization with properly structured codebooks, sample descriptions, etc. The IEA has devoted effort to setting up suitably resourced and elaborate data archives for this purpose and intends to provide archives for all its studies.

Multilevel models

In recent years statisticians have developed powerful tools, known as hierarchical or multilevel models, for studying simultaneously the between-school and between-student variation (Paterson and Goldstein, 1991; Bryk and Raudenbush, 1992). These models and associated software packages are now used extensively in so-called 'school effectiveness' studies, and in a wide variety of applications in the social and medical sciences. The scope and importance of these models can be summarized as follows.

It is well known that when carrying out statistical tests or calculating confidence intervals, account needs to be taken of the clustering of the data, which generally implies that students within a school are more alike in their test scores than students chosen from different schools. In the major international studies, this is usually done by calculating 'design effects' based upon preliminary analyses of the sample data. These design effects are estimates of the extent to which the precisions of various statistics, such as means or proportions, are inflated by such clustering. They can be used to make suitable adjustments to the statistical procedures (Skinner *et al.*, 1989). Multilevel modelling takes an essentially different approach. It attempts directly to model the hierarchical structure of the data. That is, it recognizes that a test score can be considered as the sum of contributions from each school and from each student within a school. In the simplest model, each school will have its own 'mean' score and each student a contribution to be added or subtracted from this. In more complex models, parameters such as the slope of a regression line or the difference between males and females can be allowed to vary from school to school.

Such direct modelling has several benefits. It automatically deals with the problem of accounting for the clustering of data while also providing information about differences between schools. In some cases it is the latter which is the most important focus of the analysis. For example, in a study of progress made by secondary school students in inner London (Nuttall *et al.*, 1989) the between-school variation for initially high-achieving students was found to be much greater than that for initially low-achieving students. This 'differential school effectiveness' is important when attempting to compare schools and leads to a further set of research questions. These models can easily incorporate covariates which may be defined at the level of the student, such as attitude or time spent on homework, or at the level of the school or class, such as the average ability in the class or the characteristics of the teachers.

The IEA data structures often involve complex 'matrix' designs. For example in the SSIS, there was a core test of (30) items and several 'rotated' forms in the separate areas of biology, chemistry, etc. Each student would take the core plus one or two rotated forms. While such a design reduces the burden on individual students, as opposed to completing all forms, it has meant that the resulting analysis has been difficult to carry out. Using a multilevel modelling approach such data can be handled efficiently and the approach allows full modelling of the core and all rotated forms. An example of such an analysis is given by Goldstein (1995: Chapter 4). It also obviates any practical need to collapse the separate topic areas into a single scale. The ability to focus an analysis on studying variation between schools also raises another interesting possibility, namely that comparative analyses can report differences in the extent of between-institution variation and the factors which appear to 'explain' it, rather than just differences in mean scores. Such analyses can yield valuable insights into educational structures while at the same time being less compromised by problems of translation and sampling (see below).

The problem of comparability

The continuing emphasis in international studies has been on attempts to ensure that the 'same' questions are being asked and the 'same' achievements are being measured. The difficulties associated with this have already been discussed: here I want to question whether, at least to some extent, these difficulties can be avoided by posing different prior questions.

Defining equivalence

In a strict sense the issue of whether a test, say of science knowledge, measures the same thing in two different cultures is irresolvable because there is no other external criterion which can be used to judge the issue. Rather, the problem is one of definition whereby the particular set of versions of a test have to be defined as equivalent by those responsible for producing them. Such a judgement will normally be provisional and typically contingent on satisfying reasonable criteria, for example concerning translation. Once agreement can be reached, the problem resolves itself into an empirical one of attempting to validate or dismiss the judgement. This still allows comparisons to be made, but subjects their interpretation to the general caveat that further study may cause such interpretations to be modified.

An important component of the continuing empirical validation of such judgements of equivalence will lie in the fitting of explanatory models which attempt to account for observed differences. For example, the familiarity of students with the particular type of test format used may account for some differences. It could then be argued that particular test item formats may lead to lack of real equivalence, but that this can be taken into account by adjusting for student 'exposures'. In other words, equivalence is still possible, but its definition has to be extended in an empirical fashion. One might term this 'statistical equivalence' because it is defined within the framework of a statistical model which explicitly attempts to adjust for 'nuisance' factors which are unmistakably present but which are not the focus of interest. With such a modified definition the researcher would then wish to go on to look at other factors which were associated with remaining differences.

Of course, it may not always be possible to follow such a line of reasoning. If there are countries where there is little or no variation in familiarity with different item formats, then there may be no basis for carrying out an adjustment procedure. There is also the serious problem of deciding which factors are legitimately those which can be used for adjustment. While item format familiarity might be suitable it is not so clear that a measure such as teaching methods should be used. The latter would normally be thought of as a factor of interest in its own right for explaining differences rather than in helping to define what is a fair comparison.

Second order comparisons

When modelling hierarchical structures, in addition to schools there will usually be higher level administrative or geographical units within which schools themselves are nested, with significant between-unit variation. In addition some types of schools, for example those in urban areas, may exhibit more variation than others.

Using data from the IEA second mathematics study, Goldstein (1987) found that the percentage of the total variation between schools in Japan (4%) was very much smaller than that in British Columbia (11%). This may reflect greater homogeneity of curriculum or intake achievement in Japan, but might also be a consequence of the tests used. It is strictly unnecessary to have 'equivalent' tests when carrying out second order comparisons of this kind since interest lies in the relative homogeneity of systems rather than in their absolute relationship to each other. This reinforces the argument for studying the separate components of achievement. Second order comparisons of different components could yield interesting insights into the priorities within different systems and the extent to which institutional variation was accounted for by factors such as OTL, teacher experience and so forth.

Prior achievement

In nearly all cases, large-scale international comparative studies have collected cross-sectional data, that is information on students at one point in their careers. The IEA, in the second mathematics study, collected limited longitudinal information, measuring students up to nine months apart in age. While such information can be used to estimate changes during a single school year, it is of little use for making comparisons between schools in terms of their overall contribution to students' progress. There is now a considerable literature on 'school effectiveness' studies (see, for example, Bryk and Raudenbush, 1989) which emphasizes the importance of taking account of intake achievements when students start school in order to make fair comparisons between schools. It is generally agreed by researchers that such 'value added' estimates of each school's 'effect', together with caveats about measurement relevance and reliability, are the only sound basis for comparing schools.

We can apply similar reasoning to comparisons between countries, where the purpose is to assess the relative effectiveness of educational systems. Thus, for example, comparing the achievements of ten-year-olds in mathematics may partly, or even largely, reflect pre-existing differences present when the students started school. Influences such as social background, health, parental education, etc. may all be influential. In order to isolate the contribution of the educational systems during any particular stage of education, suitable measures of student achievement prior to entry to that stage are essential. Of course, it is no easy matter to carry out long-term longitudinal studies of cohorts which would allow such analyses to be carried out, but it is difficult to see how, without such studies, any definitive conclusions can be reached. Thus, by and large, country comparisons tend to place the poorest countries behind the richer ones and this should occasion no surprise. What would be very interesting would be to know how those comparisons appeared once the initial achievements had been allowed for. Some of those countries with the lowest achieving intakes may well have secured more progress for their children. If comparisons are to be interpreted in the light of other measurements such as curriculum content or school organization then it is the value added by the schools which is the key measure.

If this argument is accepted it raises a serious problem for the usefulness of existing studies. While cross-sectional information on achievement is useful in providing a *baseline* from which to begin to draw inferences, it is only possible to begin to draw sound inferences about the impacts of educational systems and institutions from long-term longitudinal data. Similar reservations apply to other kinds of student data such as motivation and attitudes. Of course, such kinds of data are not the only kinds collected by international studies, and timely data on organization, qualifications of staff, school resources, etc. are valuable.

Conclusions

I shall attempt to summarize some conclusions and suggest directions in which I believe international comparative studies profitably could develop. First, it seems very clear that there is an important role for such studies and that the IEA is currently the most suitable vehicle to pursue them. The IEA has acted as a key forum for debating many of the relevant issues and we may expect this to continue. One of the most valuable outcomes of existing studies has been the accumulated experience gained by educationalists worldwide in the construction, analysis and interpretation of comparative data. Nevertheless, it is important that improvements are made in certain areas.

In my view, there has been an unfortunate reliance upon one-dimensional summaries of achievement test scores. In addition, the use of sophisticated statistical item response models to carry this out is an unwelcome development because it obscures too easily the true nature of what is occurring. The 'international reading scale' in the reading literacy study is a striking example. The discussion of these techniques in this chapter is an attempt to make their essential properties better understood so that informed decisions can be taken by those responsible for designing studies. Most importantly, these issues need to be well understood by governments and policy makers who are the principal providers of funds and important users of results. It seems that much of the pressure to produce simple summary comparisons has come from these latter groups and it is therefore extremely important that the issues are clear and that policy makers understand the implications of their demands. This is most appropriately done by those closely involved in international studies, and the IEA in particular could give an important lead in this. One implication is that, instead of study reports highlighting overall comparisons, they should concentrate on differential performance, properly contextualized, with discussions of any policy implications. Indeed, there is a strong case for refusing to report any comparisons in simple one-dimensional summary terms such as 'mathematics', 'science' or 'language'.

The problems of interpretation of purely cross-sectional achievement scores are legion. Future studies should begin to plan on the basis of long-term longitudinal studies encompassing, as far as possible, whole stages of education such as the elementary or secondary periods. Despite their difficulties these provide the only secure paths to proper understandings of the role of education. If this is not attempted it will become very difficult to justify large scale comparative studies if they remain solely cross-sectional.

I have referred to a number of more technical issues associated with interpretation. These are to do with equivalence across languages and cultures, the difficulties of interpretations of trends over time, the problems of properly standardizing for age and grade, the need properly to model the hierarchical structure of educational data and the importance of carrying out second order comparisons based upon the modelling of between-institution variation. In all of these areas I believe that there is important methodological work to be done which could have an importance wider than comparative studies alone. With easily available and powerful computing facilities there are no serious technical barriers to such developments

Finally, I am convinced that the existence of an organization such as the IEA with its democratic structures and enthusiastic supporters is essential. It provides the only sensible approach to making comparisons and it has often shown itself capable of responding to new issues and able to tackle difficult problems. Above all, it provides an important counterweight to the only other sources of comparative information which are based, at one extreme upon poorly designed one-off comparisons, and at the other on official government statistics. The former suffer from problems of poor controls and lack of experience, while the latter suffer from distortions related to inadequate coverage, varying definitions, and selective reporting.

Note

A more extensive version of this chapter has been published as a UNESCO report: Interpreting international comparisons of student achievement. (UNESCO Publications, 7 place du Fontenoy, Paris 75352.)

References

Beaton, A. E. and Zwick, R. (1990). *Disentangling the NAEP 1985–1986*. Princeton, NJ: Educational Testing Service.

Brislin, R. W. (1970). Back-translation for cross cultural research. *Journal of Cross-Cultural Psychology*, **1**, 185–216

Bryk, A. S. and Raudenbush, S. W. (1989). Toward a more appropriate conceptualization of research on school effects: a three level hierarchical linear model. In Bock, R. D. (ed.) *Multilevel Analysis of Educational Data*. New York: Academic Press.

Bryk, A. S. and Raudenbush, S. W. (1992). *Hierarchical Linear Models*. Newbury Park, CA: Sage.

Elley, W. B. (1992). *How in the World do Students Read?* The Hague: IEA.

Finegold, M. and Mackeracher, D. (1986). Meaning from curriculum analysis. *Journal of Research in Science Teaching*, **23**, 353–64.

Goldstein, H. (1983). Measuring changes in educational attainment over time: problems and possibilities. *Journal of Educational Measurement*, **20**, 369–77.

Goldstein, H. (1987). *Multilevel Models in Educational and Social Research*. London: Griffin; New York: Oxford University Press.

Goldstein, H. (1995). *Multilevel Statistical Models*. London: Edward Arnold; New York: Halstead Press.

Goldstein, H. and Wood, R. (1989). Five decades of item response modelling. *British Journal of Mathematical and Statistical Psychology*, **42**, 139–67.

Hambleton, R. (1992). *Translation Achievement Tests for Use in Cross-National Studies*. Vancouver: IEA International Coordinating Centre.

Hanna, G. (1993). The validity of international performance comparisons. In Niss, M. (ed.) *Investigations into Assessment in Mathematics Education*. Amsterdam: Kluwer.

Hayes, W. A. (1991). *Activities, Institutions and People. IEA Guidebook, 1991*. The Hague: IEA.

Hulin, C. (1987). A psychometric theory of evaluations of items and scale translations. *Journal of Cross-Cultural Psychology*, **18**, 115–42.

Keeves, J. P. (1992a). Scaling achievement test scores. In Keeves, J. P. (ed.). *Methodology and Measurement in International Educational Surveys*. The Hague: IEA.

Keeves, J. P. (1992b). *Learning Science in a Changing World*. The Hague: IEA.

Keeves, J. P. (ed.) (1992c). *The IEA Study in Science III; Changes in Science Education and Achievement: 1970 To 1984*. Oxford: Pergamon.

Lapointe, A. E., Mead, N. A. and Phillips, G. W. (1989). *A World of Differences*. Princeton, NJ: Educational Testing Service.

Lapointe, A. E., Mead, N. A. and Askew, J. M. (1992a). *Learning Mathematics*. Princeton, NJ: Educational Testing Service.

Lapointe, A. E., Askew, J. M. and Mead, N. A. (1992b). *Learning Science*. Princeton, NJ: Educational Testing Service.

Leung, F. K. S. (1992). *A comparison of the intended mathematics curriculum in China, Hong Kong and England and the implementation in Beijing, Hong Kong and London*. University of London: PhD thesis.

Little, A. (1978). *The Occupational and Educational Expectations of Students in Developed and Developing Countries*. IDS Research Reports, Education Report No. 3. Brighton: Institute of Development Studies.

McDonald, G. (1992). Henry and Iain—a comment on a response. *New Zealand Journal of Educational Studies*, **27**, 103–06.

Mednick, S. A. and Baert, A. E. (1981). *Prospective Longitudinal Research*. Oxford: Oxford University Press.

Nuttall, D. L., Goldstein, H., Prosser, R. and Rasbash, J. (1989). Differential School Effectiveness. *International Journal of Educational Research*, **13**, 769–76.

Paterson, L. and Goldstein, H. (1991). New statistical methods for analysing social structures: an introduction to multilevel models. *British Educational Research Journal*, **17**, 387–94.

Pelgrum, H. and Plomp, T. (1991). *The Use of Computers in Education Worldwide*. Oxford: Pergamon.

Purves, A.C. (1992). Reflections on research and assessment in written composition. *Research in the Teaching of English*, **26**, 108–22.

Robitaille, D. E. and Taylor, A. R. (1986). *A Comparative Review of Students' Achievements in the First and Second IEA Mathematics Studies*. Washington, DC: NCES.

Rosier, M. J. (1987). The second international science study. *Comparative Education Review*, **31**, 106–28.

Skinner, C. J., Holt, D. and Smith, T. M. F. (1989). *Analysis of Complex Surveys*. Chichester: Wiley.

Swain, M. (1990). Second language testing and second language acquisition: is there a conflict with traditional psychometrics? In Alatis, J. (ed.) *Georgetown University Round Table on Languages and Linguistics (GURT)*. Washington DC: Georgetown University Press.

TIMSS, TAC (1993). Summary of third international mathematics and science study, technical advisory committee meeting, Vancouver, 10–13 May 1993.

Travers, K. J. and Westbury, I. (1989). *The IEA Study of Mathematics I: Analysis of Mathematics Curricula*. Oxford: Pergamon.

Westbury, I. (1992). Comparing American and Japanese achievement: is the U.S. really a low achiever? *Educational Researcher*, **21**(5), 18–24.

Wolfe, R. G. (1989). *An Indifference To Differences: Problems With The IAEP-88 Study* (unpublished).

CHAPTER FIVE

One best system? Lessons from comparative research for assessment policy

Harold J. Noah

In this chapter two linked questions are addressed. Are there any lessons to be learned from comparative research on assessment policies and practices? Is comparative research an essential component in the making of assessment policy?

Lessons from comparative research on assessment

During the past few years I have been engaged with Max Eckstein on comparative research into assessment policies and practices in eight countries, focusing primarily on end-of-secondary-school examinations (Eckstein and Noah, 1993).[1] Four lessons, I believe, can be drawn from our study.

The primary lesson is that there is no one best system of assessment, if only because assessment goals, assumptions, policies, and practices are inextricably bound to other parts of the educational, employment, and general social framework of each country. Is it better to place the major burden of assessment in the hands of the pupils' own teachers, as in Germany? Or is it better to guarantee that none of the answer papers written in the end-of-secondary school examinations will be marked by a candidate's own teachers, as in China? All one can say is that there are advantages and disadvantages associated with each practice. Is it better to provide a lengthy set of open-book, take-home tasks in a wide range of language use and national literature, as is done in the Swedish examination at the end of secondary school? Or is it better to use the French approach of supplementing a traditional limited-time, 'no-access', written examination with a so-called *dialogue ouvert*, an oral examination in which candidates are asked to discuss with the examiner various aspects of the literary works they have listed as having previously read? Or is the 'set book' approach in English literature used by examiners in England more desirable? Merely to ask such questions is to suggest they are inherently unanswerable in the form given.

A second lesson of comparative study is that tests, examinations, and other modes of assessment can be powerful levers of change. Although existing forms of assessment primarily reflect the established ways of the schools, changes in assessment content, standards and procedures can alter the way teachers and pupils go about their business. When the Swedes abandoned traditional 'one-shot', set-piece examinations in favour of continuous assessment and in-course tests and examinations, some profound changes in the structure of upper secondary education and pedagogical practice became possible. China's on, off, and on again switches of university entrance examinations promoted radical swings in the operation of secondary schools. Currently in England the Conservative government is persisting in its programme of introducing regular national testing of school children in the major school subjects. The hope is to raise the achievement levels of many children who, the government claims, have been short-changed by the traditional autonomy of each school to set its own curriculum and standards. Even in the United States, where the notion of a national assessment of individual achievement measured against nationally recognized standards would have been unthinkable a dozen years ago, the unthinkable may yet come to pass. As in England, the expectation is that a major change in assessment will change the behaviour of all connected with schooling.

A further lesson from comparative study relates to the control of examinations, particularly national external examinations. These differ from country to country according to the location of authority and responsibility. Arrangements for external examinations normally follow the pattern of control of the educational system as a whole. We expect that in nations where the school system is directed from the centre, external examinations are also centrally directed, and where regional and local authorities enjoy substantial autonomy in school affairs, there will be found substantial regional/local control over examinations, too. France is a clear example of the former, Germany of the latter.

Less obvious is the evidence that even in the most centralized systems, some degree of devolution of the examining and assessment process is the rule. Regional or local decisions determine the choice of questions and the grading of answer papers in France, perhaps not quite as much as they do in Germany, but nevertheless quite substantially. The same is true in China, which in formal terms runs a highly centralized system. This was also certainly the practice in the former Soviet Union, where what was supposed to be centrally coordinated over the entire country, by Communist Party oversight, in fact allowed for a wholly unannounced degree of decision making at the local, and even individual school level.

Conversely, decentralized systems can and do arrange for effective national coordination within their federal structure. In Germany, although assessment policy (the *Abitur* especially) is in the hands of each *Land*, provision for coordination via the Standing Conference of Ministers of Education (the KMK) is extremely well developed, and has made possible a marked reduction in variations in assessment practice among the *Länder*. While other federally organized countries, such as the United States and Canada, are far from achieving the German degree of coordination, they too are not without mechanisms for reining in centrifugal tendencies in educational matters. Canada has an analogue of the German KMK in the Council of Ministers of Education, Canada. In the United States, the Council of Chief State School Officers and the federal Department of Education in Washington, DC, have been increasingly willing and able to influence the states in the direction of greater attention to standards of achievement and better assessment practices. The council has been a leading force in the extension of the National Assessment of Education Progress, and the Department of Education has been very active in promoting the development of national standards of educational achievement.

Last, we should note that the comparative perspective offers strong support for the US and Swedish practice of using tests and examinations more for purposes of 'selecting in' than for 'selecting out'. Combined with the provision of multiple opportunities for demonstrating talent, actual or potential, the US and Swedish approach is in sharp contrast with the Chinese and Japanese practice of using examination results as an effective, economical and socially defensible way of cutting down the overwhelming number of candidates to fit the limited number of publicly-funded study places. The American and Swedish approach weakens incentives to study hard and achieve academically, but gains by increasing the chance that those with talent are not eliminated prematurely from subsequent educational opportunities. The approach in China, Japan, and many other countries is based on the assumption that success in formal written examinations is the most expedient way to identify talent and/or industriousness and to award places in state-subsidized education. A grave weakness of this approach is that examination success is not only (or even mainly) the result of innate talent. The quantity and quality of the resources parents devote to their children's upbringing, including the quality of the school their children attend, is also very important. Highly competitive selecting-out examinations tend to favour candidates from better-off families and too often fail to uncover much of the reservoir of talent in the rest of the population.

Is comparative study an essential component in the making of assessment policy?

The 'world systems' approach to comparative education, associated primarily with the work of Meyer and his colleagues at Stanford University, emphasizes the commonalities rather than the differences among contemporary education systems (Meyer *et al.*, 1979; Thomas *et al.*, 1987). As evidence for their perception, these scholars point to the standardization across the globe of such characteristics as: the tripartite division of formal schooling into elementary, secondary and higher levels; commonalities in school curricula; and similarities in the training, appointment, and conditions of work of teachers. To this list can be added a shared global concern with assessment policies, practices, and outcomes, as well as some fairly strong signs that assessment practices are tending to converge.[2]

The commonalities have arisen because governments share so many of the same aims for their schools: universal coverage; education for citizenship and employment; and promotion of the taste for life-long learning. Comparative study, too, has played a part in spreading more uniform practices around the globe, as its practitioners go about their work of observation, reporting, and advising.

Although not every nation paid attention to what was happening elsewhere in the world when they established their educational systems, many did. Alongside attention to the ways in which foreign countries organized their education systems—the structure and content of curricula and methods of training teachers—assessment practices came in for particular scrutiny and copying.

For example, China threw out its venerable, 2000-year-old examination system in 1904, and instituted a set of qualifying examinations for government service based entirely on Western models, as part of the belated attempt to face up to the challenge of European, US, and Japanese incursions (Hu, 1984).

In the last quarter of the nineteenth century Japan sent commissions to Europe and the United States to study educational policies and practices. Their reports formed the basis for a sequence of trial-and-error reconstructions of schools and higher education. So closely did the Japanese reformers wish to conform to a European system that at one point in the early Meiji Restoration period, higher education courses were given in German! But the Japanese were nothing if not eclectic, trying first one and then another foreign example before arriving at the turn of the century at a reasonably stable melding of practices borrowed from abroad with their own traditions. Among the borrowed practices one of the most influential was the introduction of highly competitive, demanding entrance examinations at each transition point to the next stage of education, a policy which has persisted down to the present (Amano, 1990).

As secondary education in the Russian Empire was expanded in the 1870s, the German *Gymnasium* became the preferred model of the Tsarist administration. In particular, the examination at the end of secondary school was modelled on the contemporary German *Abitur*, which emphasized Latin classics and mathematics and gave pride of place to oral rather than written examination.

Phases in the development of education in China, Japan and Russia are examples of what may be termed voluntary borrowing from abroad, but this mode of spreading models of education around the world was complemented by another. The nineteenth and twentieth centuries have been marked by a vast amount of involuntary transfer of foreign models— primarily the many examples of imposition on subject peoples of the educational models of colonizing or imperial powers. While this hardly falls under the heading of 'borrowing', the consequences were much the same as voluntary adoptions and adaptations—certainly many of the involuntary implantations have persisted as long, if not longer, than the voluntary adoptions. The colonizing powers were represented not only by their missionaries, traders, planters, armies, government officials, and school teachers, but also by their examination systems. The United Kingdom, France, Germany, Portugal, and the United States made the greatest impact on the educational scene in their colonies and dependencies. Indeed, although formal political independence has long been achieved in African countries, thousands of anglophone students in Africa continue to take examinations constructed and graded in Cambridge, England, just as their Francophone peers take examinations modelled closely on the French *baccalauréat*.

Although they did not arrive as colonizers in the aftermath of World War II, the four Allied occupation administrations ensconced themselves in Germany, as did the Soviet forces in Eastern Europe, and the US military in Japan. In West Germany the innovations brought to education by the British, French and North Americans did not long survive the restoration of West German sovereignty. The North Americans in Japan had somewhat longer lasting impact on policies and practices, but not especially so in the realm of assessment and examinations. On the other hand, Soviet innovations have been deeply implanted in their Eastern European sphere of influence, in all major aspects of education. As Soviet power crumbled there in the late 1980s, many of the structural changes in education brought about by Soviet 'advisers' remained more or less intact, East Germany being perhaps the exception to this generalization. Some of the hallmarks of the Soviet assessment system, such as the five-point grading system and the reliance on oral examinations, appear to have survived the departure of Soviet political and military power.

Voluntary and involuntary borrowing and adaptation of assessment approaches have brought both advantages and disadvantages. Japan's experience is most instructive in this respect, if only because Japan appears to have been so successful in choosing those aspects of foreign models that could be adapted to Japanese goals and modes of operation. Reformers found that citing the experience of a powerful foreign country lent weight to arguments for change in the desired direction. Comparative reports have often been used in this way, either to support official policies for reform or to undermine the case for remaining with the status quo. Once it was decided to go ahead with the introduction of particular foreign practices, scholars and administrators from that country could be brought in to advise on and administer the new arrangements. In addition, Japanese could be sent to the source country to learn the new approach on the spot, substantially shortening the time between initiation and full implementation of the new practice.

But while Japan may have been notably competent at the business of adapting foreign models to local needs, this has not been the typical experience of borrowing from abroad. Indeed, the educational landscape is littered with the ruins of importations that signally failed to live up to the promises made on their behalf. And even in Japan, it could be argued that the borrowing and adaptation of Western models of assessment have simply gone too far, that the Japanese have fallen into the trap too often associated with borrowing from abroad—an excess of enthusiasm for the borrowed practice. Assessment and examinations in particular are arguably both necessary and desirable, but as with so many such things, only when used in moderation. There are those in Japan, as well as abroad, who deplore the way in which secondary education in Japan has become so firmly fixed on the business of passing examinations (US Department of Education, 1987).

National education systems are also capable of being sturdily resistant to foreign imports. Although the Japanese have adopted US-style multiple-choice, machine-scorable tests and although in China these techniques are gaining ground quite rapidly, examinations in England, France, and Germany remain firmly rooted in their traditional extended answer, open-ended formats. Each has borrowed remarkably little from the other two in the development of its school system in general and of its systems of assessment in particular. There is certainly a good deal of rhetoric in England and France about the virtues of the German dual system of apprenticeship training and continued secondary schooling; in France and Germany about the desirability of the (alleged) emphasis in English education of character training; and in England and Germany about the advantages and disadvantages of the French tradition of centralized, self-conscious, *national* education. But with relatively few exceptions, none of them with any lasting significance, comparative study and foreign example made little impression on the development of educational systems and modes of assessment in the three countries. In these matters, as in so much else in the social sphere, national idiosyncrasy is, and has been, the rule.[3]

Notes

1. The eight countries studied were China, England/Wales, France, Germany, Japan, Sweden, the former Soviet Union, and the United States of America.
2. Systems that have centralized control of examinations are incorporating more elements of devolution (France, China, and Sweden), while decentralized systems are moving to more central control and the encouragement of uniformity (England/Wales, the United States). Examinations are everywhere incorporating more school subjects than before, and candidates are being offered more choice of subjects and levels of difficulty than before. Examination techniques are also tending to converge: traditionally open-ended, extended answer questions are being complemented with multiple-choice items, while those systems which relied primarily on multiple-choice formats are beginning to include some open-ended questions. Everywhere, too, the fraction of the age-group taking examinations has risen, so that examinations that were once quite exclusive are now mass enterprises.
3. One can of course look back in English education history to the mid-seventeenth century Commonwealth, when Comenius came to England from Bohemia at Oliver Cromwell's invitation. His task was to help establish a national system of education and to improve pedagogy. But the changes made on his advice were short-lived, extinguished immediately upon the Stuart Restoration. Also, although the school practices of nineteenth-century Prussia were highly praised by French and English observers (Victor Cousin and Matthew Arnold were perhaps the most prominent), by and large the French and English systems developed independently of any general adoption of Prussian principles and practice. On the other hand, US higher education was greatly influenced first by importations directly from Scotland, and then from Germany.

References

Amano, I. (1990). *Education and Examination in Modern Japan*. Tokyo: University of Tokyo Press.

Eckstein, M. A. and Noah, H. J. (1993). *Secondary School Examinations: International Perspectives on Policy and Practice*. New Haven: Yale University Press.

Hu, C. T. (1984). The historical background: examinations and control in pre-modern China. *Comparative Education*, **20**, 7–26.

Meyer, J. W. *et al*. (1979). The world educational revolution, 1950–1970. In Meyer, J. W. and Hannan, M. T. (eds) *National Development and the World System: Educational, Economic, and Political Change, 1950–1970*. Chicago: University of Chicago Press.

Thomas, G. M. *et al*. (1987). Institutional Structure: Constituting State, Society, and the Individual. Newbury Park: Sage Publications.

US Department of Education (1987). *Japanese Education Today*. Washington, DC: US Government Printing Office.

Part two

COUNTRY CASE STUDIES: SYSTEMS IN TRANSITION— EXPERIMENTS IN LARGE-SCALE REFORM

The major focus of this collection is, by design, the way assessment is developing and changing worldwide. It is true that at any point in the last 150 years one could identify a country where major changes were underway in assessment and examination systems and their impact on society. Some of the characteristics of national systems which we see as fixed and defining—such as the United States' reliance on multiple-choice tests devised by huge independent agencies—are actually of quite recent origin. Nonetheless, there is probably more change underway now, worldwide, than at any time since the major European countries of the nineteenth century adopted formal examinations for universities, the civil service, and a range of licensed occupations.

The following four chapters focus on countries chosen because of their involvement in large-scale assessment reforms. They include the world's richest country, the United States, and its most populous, China. England and Wales attempted the most ambitious experiments in assessment-based reform of the late 1980s and early 1990s; while Chile has gone furthest in implementing the type of centralized government-run system of monitoring quality currently advocated by the World Bank and UNESCO.

In all our examples, there is evidently an appetite for more and more information, and an optimism about the ability of administrators, teachers and parents to absorb and use this information in a productive fashion. To a considerable degree this appetite, and this optimism, are a product of technology. New systems for collecting and analysing data make it possible to collect more complex data on a large scale, where previously resource constraints made it impossible to do this for more than the most superficial measures. New technology also seems to promise something even more desirable: that large scale assessment data can be analysed fast enough actually to be of use to the education system. All too often in the past, survey results have appeared only years after collection, when not only the students but often their teachers and the commissioning bureaucrats and politicians had long since moved on. In such circumstances, a considerable scepticism about the worth of monitoring (or accountability) systems was hardly surprising.

It remains to be seen whether technical advances will translate into genuine improvements in the quality and efficiency of education. The overambitiousness of English reforms (Chapter 8) provides a note of caution; but certainly technical changes now offer the possibility, worldwide, of systems which, even ten years ago, were beyond the reach of even the richest countries.

However, while the case studies which follow all chart reforms which draw on technical advances, they also underline the variety of these same reform processes. In Chile, the motivating force is the desire to monitor quality, focus state resources, and achieve tighter central oversight in the interests of efficiency. Raising the lowest levels of basic education is a political imperative for which international loans have been provided; and the accompanying reliance on multiple-choice testing of the US sort is not, apparently, an issue. In the United States, by contrast, the effect of such tests on the curriculum is one of the major issues fuelling the reform movement. Here, the effect of assessment on the nature of learning is far more important and evident than in our other three case-study countries. Moreover, the discontent of North Americans with the current way in which assessment affects curriculum and learning is related as much to the perceived failure of their system to reach acceptable 'high' and 'average' levels as with quality at the bottom end.

There is also no great consistency in the relationship between active reform and the allocation of power and decision making in the education system. In England, reform was designed to shift power away from local education authorities and towards central government on the one hand, and parents on the other. In China, by contrast, the shift is away from the centre, with new powers being accrued by the provinces and municipalities. In the United States, activity centres on 'professional' test constructors, and a similar pattern can be found in France. However, a broader look indicates that in other countries— as in Australia and Sweden—assessment reforms designed to broaden the curriculum have vested increased power in teacher-assessors.

Finally, it is important to emphasize that, while assessment reform may be unusually widespread and far-reaching, it is neither universal nor equally ambitious. France, for example, engages in universal 'assessment for monitoring' with just two age groups compared to four in England. Moreover, while there have been moves to more continuous assessment and different questions intended to promote types of learning, change has been far more gradual and modest than attempted by English or envisaged by North American reformers. In some countries, moreover, the speed of change is glacial or apparently altogether absent. Germany and Japan are two of the late twentieth century's most successful economies, with highly distinctive and different education systems. In Germany, assessment is devised by the teachers; in Japan, psychometric tests (largely multiple-choice) are the rule. Both systems boast high levels of achievement across the board; and in both—though especially Japan—the importance of school results for future success in life creates huge pressures for students and also explains their general motivation and hard work. In neither country is major reform under serious consideration; we would suggest because of the general view that the education system is working well by international standards. The conventional wisdom that education matters more and more to economic success is one reason why we would predict continuing large-scale reform efforts around the world. It may nonetheless help block reforms elsewhere—with all the good and the bad that an absence of change implies.

CHAPTER SIX

A turning point for assessment: reform movements in the United States

George F. Madaus and Anastasia E. Raczek

Testing in the United States is at a turning point, and there are many directions in which it may turn. Will testing become the prevailing symbol for education in the United States—a universal, high-stakes, technology the results of which will be used to decide whether students receive a high-school diploma, go on to college, or secure a well-paying job? Or will testing take a back seat, and serve as a complement to well defined and appropriate school curricula? Important policy decisions currently being debated will influence which way the testing tide will turn.

In this chapter we consider first the extent of standardized testing in the United States. Standardized testing in US schools has grown over the past few decades in a manner that can only be described as exponential. Several indicators of this increase in testing are discussed, including mounting sales of tests and test-related services as well as growth in the number of references to testing in the educational literature. Next, changes in the way test results are used and the influence of testing in the classroom are described. We consider specific social forces that have acted to transform standardized testing into a critical component of US education. Finally, we describe current proposals for a national assessment system. We conclude by delineating issues that must be resolved if a national assessment system is to be established in the United States.

The US educational system is enormous. During the 1989/90 school year, it included 15,367 school districts comprised of approximately 109,232 schools. In those schools were over 45 million students; 60% in elementary schools (Grades K–8), 34% in secondary schools, and about 6% in combined schools (Snyder, 1991). One activity common to all these students is sitting for standardized testing.

While testing has been used as a policy tool in US education at the local level since at least the 1840s (Tyack and Hansot, 1982; Massachusetts Historical Society Documents, 1845–46; White, 1888; Madaus and Kellaghan, 1992; Madaus and Tan, 1993), the nature and magnitude of test use in the United States has changed since World War II. Over the past fifty years, the various uses to which educational testing is put have proliferated. These uses include: guidance; creativity testing; use in the mastery learning movement; test scores used as school effectiveness indicators; and use in the criterion/curriculum referenced testing movement (Madaus and Tan, 1993). Additionally, the development of test standards and codes of testing by professional organizations and the establishment of the National Merit Scholarship Corporation were important outgrowths of testing. However, a significant, and recent, development in US educational testing is the growth in the use of standardized tests as policy tools for educational reform.

In addition to teacher-administered assessments covering class material, US students are subject to many standardized tests as part of state-mandated testing programmes and school district testing programmes. Table 6.1 summarizes the number of tests US children take.

Special needs and bilingual students often experience additional standardized testing (National Commission on Testing and Public Policy, 1990), and millions of secondary students take one or more college admissions tests. Considering all these sources, anywhere from 143 million to 395 million tests are administered to US students annually (Haney *et al.*, 1993). Calculation of the number of tests administered annually can differ depending on estimates of the number of students assessed and whether a multitest battery administered in one sitting is counted as one exam or several.

The growth in educational assessment

Several indicators register the growth in testing during the last 50 years, including recent increases in the number of state-mandated testing programmes, sales figures from test publishers, increasing numbers of references to testing in the education literature, and comments from stock-market analysts on testing as a growth industry (Haney *et al.*, 1993).

The level of standardized testing in elementary and secondary schools has grown enormously over the past three decades. The number of states authorizing minimum competency programmes and assessment programmes rose dramatically, beginning in the mid-1960s. From 1950 to the mid-1960s, fewer than five states authorized such tests. By 1990, *every* state mandated some kind of testing programme (see Fig. 6.1). Independent non-profit or for-profit companies are responsible for collecting most standardized assessment data in the United States.

Table 6.1 Numbers of educational tests given annually in the late 1980s

State mandated testing programmes	Low estimate	High estimate
Number of students tested	11,000,000	14,300,000
Number of tests per student	3	5
Subtotal	33,000,000	71,500,000
School district testing programmes		
Number of achievement tests	82,668,966	248,006,898
Number of ability tests	2,952,463	23,619,705
Subtotal	85,621,429	271,626,602
Special populations		
Number of tests for special education students	7,920,000	19,800,000
Number of tests given to bilingual students	3,600,000	10,800,000
Subtotal	11,520,000	30,600,000
College admissions testing		
Subtotal	13,034,318	21,759,548
Total	143,175,747	395,486,150

Sources: Haney *et al.* (1993); Office of Technology Assessment (1987).

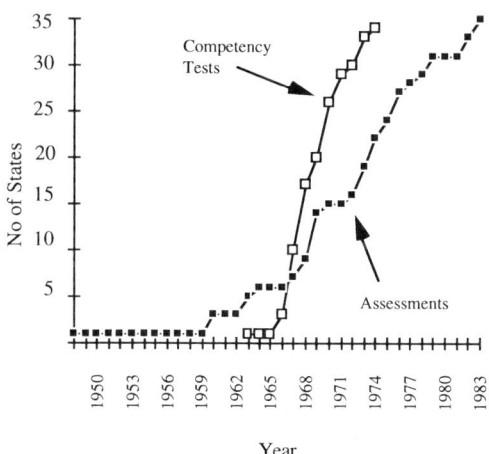

Figure 6.1 Numbers of states authorizing minimum competency testing and assessment programmes (source: Haney *et al.*, 1993).

Growth in test sales

A second indicator of growth in testing is the increase in reported dollar volume of sales of tests and testing services at the elementary and secondary level. Sales figures for standardized tests in the elementary and secondary market from 1955 through 1991 are shown in Fig. 6.2. These figures are adjusted for inflation using the 1988 Consumer Price Index (*The Bowker Annual*, 1970–1991; Haney *et al.*, 1993). Figure 6.2 shows a tremendous rate of increase in the real dollar volume of sales of tests and testing services from 1955 to 1987. Testing in the United States is a commercial industry—and a big business. These increased revenues of the testing industry are due more to the increased volume of testing than to increases in the costs of tests or test scoring services (Haney, Madaus and Lyons, 1993). It should be noted that sales data from the *Bowker Annual* do not include revenues from some test scoring services, sales figures from companies that are subcontracted to develop standardized tests but do not publish such exams, and revenues of the ETS and ACT—two of the largest testing firms in the United States.

The US testing marketplace is dominated by seven companies, the two largest of which account for almost 70% of the $690 million in sales by the top seven. The ETS is a powerful force in the college and graduate admissions testing market and is usually considered to be the largest testing company in the country (Buros, 1978; Haney *et al.*, 1993). Total revenues of ETS showed a marked increase from $35 million in 1970 to $310 million in 1991, a compound annual growth rate of 11%. National Computer Systems (NCS), a public company, has a near-monopoly on scoring scan-

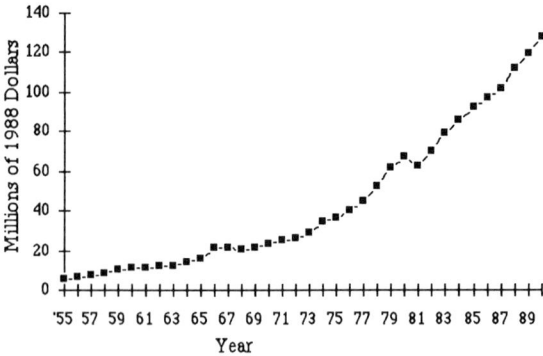

Figure 6.2 Standardized test sales, 1955–90, in millions of 1988 dollars (source: Haney *et al.*, 1993).

nable standardized tests and is a leader in the sales of scanning machines. CTB/McGraw-Hill is probably the major publisher of elementary and secondary group tests, with the Psychological Corporation/Harcourt Brace Jovanovich a close second.

Growth in references to testing

An interesting indirect indicator of the growth in testing developed by Haney and Madaus (1986) charted the number of citations in the *Education Index* under the rubric 'tests and scales' (as indicated by number of column inches) from 1930 through to 1988. For comparative purposes, citations under the 'curriculum' rubric were also charted. In terms of the sheer *numbers* of articles under the testing rubrics found in *Education Index*, there were 179 articles in the edition covering the years 1941–44; by the 1990–91 edition, the number had swelled to 728; the number of articles peaked between July 1984 and June 1985 at 1154 titles (Haney *et al.*, 1993).

Figure 6.3 shows that the average annual number of column inches devoted to citations concerning curriculum increased only modestly over the last 62 years, from 50 to 100 inches per year in the 1930s and 1940s to only 100 to 150 in recent years. In contrast, column inches devoted to tests and scales have increased dramatically, from only 10 to 30 in the 1930s and 1940s to well over 300 in 1990–91. While this index may be somewhat crude, the data certainly indicate that the prominence of testing in US education literature has grown dramatically, particularly since the mid-1960s.

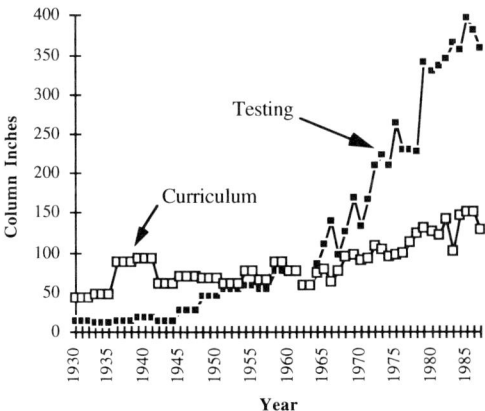

Figure 6.3   Education index listings under testing and curriculum (source: Haney *et al.*, 1993).

Changes in test use

Not only has testing grown in the United States; the uses to which test results are put have also changed dramatically. The National Commission on Testing and Public Policy (NCTPP) (1990) noted that the growth in testing since the 1950s accompanied a trend of greater reliance on test results to make critical decisions about children, such as:

• entry to and exit from kindergarten
• promotion from grade to grade
• placement in remedial programmes
• graduation from high school.

Further, there was a dramatic increase in the use of students' scores to hold school systems, administrators, and teachers accountable (Madaus and Tan, 1993).

More recently, in 1992, the National Council on Education Standards and Testing (NCEST) endorsed the use of assessments to monitor individual and system progress toward the national education standards and to:

• exemplify for students, parents, and teachers the kinds and levels of achievement that should be expected
• improve classroom instruction and improve the learning outcomes for all students
• inform students, parents, and teachers about student progress toward the standards
• measure and hold students, schools, districts, states, and the nation accountable for educational performance
• assist in education programme decisions to be made by policy makers (NCEST, 1992).

Thus, not only has the volume of testing increased, but testing has become a high-stakes policy tool looming more ominously over the lives of many educators and children, influencing what they teach and how, and what is learned and how (NCTPP, 1990).

Effects of standardized tests on the curriculum

What effect has the growth in large-scale, high-stakes assessment had on the school curriculum? Most educators agree that tests *do* affect curricula; while some claim the curricular influence of tests is a positive one, benefiting teaching and learning, others assert it has a negative impact on schooling (Kreitzer and Madaus, 1993). One proponent of high-stakes tests states that these exams 'serve as a powerful curricular magnet' that causes teachers to 'focus a significant portion of their instructional activities on the knowledge and skills assessed by such tests' (Popham, 1987: 360). On the other hand, opponents of measurement-driven instruction argue that too many high-stakes tests cause teachers to teach to the test and neglect other important, non-tested skills. Standardized tests can narrow the school curriculum to the point that instruction becomes mostly test practice (Kreitzer and Madaus, 1993). One study found that exams have a strong influence on curriculum especially when high-stakes are associated with standardized tests. For example, when school administrators exert a great deal of pressure on teachers to do well because test results will be used for accountability purposes, teachers are more likely to teach to the test (West and Viator, 1992). The problem of test-driven curriculum becomes especially important in light of the increase in standardized testing as described above; if students are asked to take more and more high-stakes tests, more and more classroom time may be spent on test preparation.

Reasons for the growth in testing

A variety of social forces have contributed to the steady growth in standardized testing in the United States. In particular, four broad social forces operating during the past 50 years help explain the transformation of testing. These forces include public dissatisfaction with the quality of education, increases in federal and state legislation that mandates testing, a shift in attention from educational *inputs* to educational *outputs*, and the increased bureaucratization of both schooling and society (Haney *et al.*, 1993). These broad forces do not work independently. For example, a specific episode of public dissatisfaction with education often leads to legislation mandating new tests, which in turn focuses public attention on outcomes of schooling such as test scores; and legislation and testing seems to increase the bureaucratization of education.

Dissatisfaction and efforts at reform

Four major cycles of educational dissatisfaction and reform in the United States have spurred this growth in standardized testing during the past 30 years, either by using test results to inform policy makers about the state of education or as a policy mechanism aimed at transforming education. First, the 1960s civil rights movement spurred efforts to improve education for the disadvantaged. However, it also brought increased attention to standardized test results in two very different ways. Advocacy groups used test results of economically disadvantaged students to support proposals for protecting the civil rights of minorities and improving the educational opportunities of disadvantaged children. On the other hand, standardized tests became the yardstick by which the success of new compensatory education programmes such as Head Start and Title I was judged (Madaus, 1985). The civil rights and compensatory education movements also led to legislation which had a serious impact on testing and the testing industry (Madaus and Tan, 1993).

A second cycle of public dissatisfaction and reform is marked by the furore over the 1970s national declines in SAT (pre-college) scores. The national average verbal SAT declined 50 points and the math SAT declined 30 points from 1963 to 1977; this decline sparked debate about the quality of education for high-school students in public schools. These declines also contributed to the rise of the minimum competency testing (MCT) movement (Haney *et al.*, 1993).

Test results were again used in the 1980s to call for reforms. Several reform reports released in the 1980s—most notably *A Nation At Risk* (National Commission on Excellence in Education, 1983) and *High School: A Report on Secondary Education in America* (Boyer, 1983)—used data from the National Assessment of Educational Progress (NAEP), IEA surveys (see Chapter 4), the SAT and other batteries to describe the state of secondary education. These reports used test score data to argue that the nation's schools were failing. These reform efforts rekindled concerns over US competitiveness, now economic rather than military. Most of the reports also called for testing as a policy mechanism that would remedy the ills disclosed by testing in the first place.

This decade has continued the tradition of using test results as an instrument of educational reform. The early 1990s featured a renewed concern with US economic competitiveness and what was perceived as poor performance in mathematics and science of US students compared to those in other countries. In response, President Bush and the nation's governors (led by then-Governor Clinton) agreed on six goals for the reform of US education. 'America 2000' incorporated those goals and called for the establishment of 'world-class standards' and a national testing programme to monitor progress toward the goals and the attainment of those standards (US Department of Education, 1991). The NCEST was created in June, 1991 to provide Congress with advice on the desirability and feasibility of national standards and testing in education. Some argue that the feasibility of creating a national testing system was never adequately addressed by NCEST (Koretz *et al.*, 1992). Nonetheless, in its seminal report NCEST maintained that 'standards and assessments linked to world class standards can become the cornerstone of the fundamental, systemic reform necessary to improve schools' (NCEST, 1992: 5). Testing, albeit labelled 'authentic' assessment, was once again touted as the policy tool-of-choice to hold schools and individuals accountable and to re-energize US schools.

Legislation and testing

The US school system is subject to a complex web of controls, including federal, state, and local entities. Federal controls include legislation mandating educational programmes and activities, Supreme Court decisions controlling how schools will educate, and Department of Education policy. State controls include state Education Agency policy, state court decisions, and state Constitutional provisions. Finally, local controls include school-board policy, local charters, administrative leadership, and parent–teacher organizations (Haney *et al.*, 1993).

A direct influence on the prominence of standardized tests has been federal legislation mandating testing programmes. Five major pieces of federal legislation enacted over the past 30 years have contributed to increased testing (Haney *et al.*, 1993).

1. The *1964 Civil Rights Act*, brought increased attention to testing. Title VII mandated non-discrimination in employment by reason of race, sex or national origin, and it was widely used to challenge employment testing (Haney *et al.*, 1993). The 1964 Civil Rights Act also mandated the Coleman Report, which helped to shift the definition of equal opportunity in schooling from educational inputs to outcomes, as measured by standardized tests.

2. The NAEP, a regularly administered Congressionally mandated assessment programme, was implemented through legislation in 1963, and represented the first federal funding for collection of nationally representative test data on a continuing basis (Greenbaum *et al.*, 1977). NAEP is designed to provide achievement data about what students in US schools know and can do for the nation as a whole, as well as for subpopulations of students (Beaton and Zwick, 1992). In addition to traditional multiple-choice items, recent NAEP administrations include open-ended questions that allow students to produce their own answers. NAEP also contributed to the shift to focusing on outcomes when considering educational quality; NAEP data were used in the 1980s and 1990s to support calls for educational reform.

   NAEP has also contributed to the development of testing technology in scaling, sampling, and reporting results (Madaus and Tan, 1993). Because NAEP is not designed to provide information about individual students or individual schools, its administration design is complex. A multiple matrix sampling procedure allows for estimation of population characteristics from test results, while it avoids placing too much of a burden on students and schools (Beaton and Zwick, 1992). Through the recent Trial State Assessments (TSA), NAEP has administered tests that allow student performance data to be reported at the individual state level as well as national level. It has been proposed that NAEP be extended downward to district, school, and even individual levels (National Assessment Governing Board, 1990; US Department of Education, 1991). These later proposals have the potential to change the character of NAEP as an independent, valid indicator of the nation's educational progress, and need to be watched closely (Haney and Madaus, 1992).

3. The *Elementary and Secondary Education Act* (ESEA) of 1965 mandated that the federal government provide assistance to local education agencies serving low-income families. State education departments were also provided with funds for measuring state-wide student achievement and evaluating Title I (now Chapter I) programmes for the disadvantaged and other interventions such as Head Start (Title II of the Economic Opportunity Act [ECO] of 1964) through standardized test results. Standardized test results became the measure by which Title I programmes were evaluated. ESEA also provided money under Title IV for the training of personnel conducting tests and measures.

4. The *Education For All Handicapped Children Act* of 1975 (renamed the Individuals with Disabilities Education Act in 1990) contributed to the increase in testing by promoting tests and evaluations to determine placement, assess areas of need, and evaluate effectiveness of individual educational plans (IEPs) mandated for special needs children (Haney and Madaus, 1992). 1986 amendments to this act extended the right to 'free and appropriate' education for children with disabilities to include preschoolers aged three to five.

5. The *Augustus F. Hawkins, Robert T. Stafford Elementary and Secondary School Improvement Amendments of 1988* contained the first-of-their-kind provisions for a federal test of individual students. It authorized the secretary of education 'to approve comprehensive tests of academic excellence or to develop such a test where commercially unavailable, to be administered to identify outstanding students who are in the eleventh grade of public and private secondary schools' (P. L. 100–297: 102 STAT. 247–48). The bill also authorized the secretary to award certificates to students who scored at a sufficiently high level on such tests. These provisions have never been implemented, however.

Focus on outcomes of schooling

The increasing use of tests as both a diagnosis for educational ills and a prescription for reform, as well as escalating legislative attention to testing, points to a fundamental shift in the way in which people regard the quality of schools. For most of this century, US educational reformers such as

James Bryant Conant (1961) and Francis Keppel (1965) viewed school quality in terms of a range of resources, facilities and conditions (Haney *et al.*, 1993). School quality depended on *inputs,* such as the quality of the physical plant, characteristics of teachers, and school finance. The 1966 release of the *Equality of Education Opportunity* report (EEOR), or Coleman Report, found that 'schools bring little to bear on a child's achievement that is independent of his background and general social context' (Coleman, *et al.*, 1966). This widely publicized finding dramatically shifted the focus of discussion about equality of opportunity away from inputs to the *outcomes* of schooling (Madaus and Tan, 1993). It should be noted that the EEOR report's major finding was widely debated, and the data closely scrutinized by both researchers and educational policy makers (Haney *et al.*, 1993). The shift away from school resources toward school outputs measured by tests of academic achievement clearly contributed to the prominence of testing as a policy tool of accountability from the 1960s to the present.

Bureaucratization of education

Finally, testing has served to change US education over the past 50 years by making it increasingly bureaucratic (Hall, 1977; Wise, 1979; Haney *et al.*, 1993). Control of modern schools has become increasingly centralized— many school districts have consolidated to form fewer, larger systems, state education agencies have become more influential than local agencies in many states, and federal involvement in education has become increasingly prominent (Haney *et al.*, 1993). The growth in testing fits well with such bureaucratization. For example, tests can provide a means for categorizing people and institutions according to generalizable rules. Scores for individual students can be aggregated to describe performance of larger units, such as schools, districts, or states. Also, testing is an efficient, administratively convenient, mechanism for policy makers to not only measure performance of students and educators, but also to affix important rewards or sanctions to that performance (Madaus and Tan, 1993).

The costs of testing

The growth in educational assessment may have also evolved from the notion that since standardized testing is so inexpensive and holds the promise of providing an 'objective' measure of how schools are performing, it always pays to increase the level of testing. However, such reasoning disregards the substantial *indirect* costs associated with testing and does not specify the benefits associated with increments in educational testing. Haney *et al.*,

(1993) estimated that state and local governments invest as much as $20 *billion* annually in standardized testing programmes; $86 million to $4.7 billion annually in state assessment programmes, and between $225 million and $18 billion in district testing programmes (in 1988 dollars).[1] This estimate consists of several components: the *direct* cost of test administration per student per hour of testing time; administrative or *transaction* costs that are related only indirectly to the achievement of a policy goal but are nonetheless still paid in cash from public or private bourses; and by far the larger indirect cost, the *opportunity* cost of the time that teachers, administrators and students devote to standardized testing in the nation's schools. It can be difficult to ascertain the value of time that teachers, administrators and students spend on an activity. In general, however, the appropriate value of time devoted to testing is the value of the next best alternative activity to testing. For staff, the most accessible measure is their wage rate. It is important to remember that, although the value may be difficult to calculate, student time is also worth something and an hour spent taking a test is one less hour students can spend in class learning.

Since total national expenditure on elementary and secondary education in the United States during 1987–88 was about $169.7 billion (Snyder, 1991) these testing cost estimates make it clear that testing is not simply a cheap activity for monitoring educational outcomes.

Current trends in assessment

Testing continues to be used as a policy tool aimed at educational reform. Much activity in US educational assessment today, however, focuses on the *method* of testing. Methods for examining student achievement in the United States have been modified over time. The predominant mode for administering examinations has changed from the oral mode to written essay to short answers to multiple choice. The most prevalent method today is via machine-scorable answer sheets, which became possible to do after the introduction of optical scanners in the mid-1970s (Haney *et al.*, 1993). These shifts in examination method have resulted in examinations that are more efficient and manageable, standardized, easily administered, objective, reliable, comparable, and inexpensive (Madaus, 1993a).

At the beginning of the 1990s, a reaction against the administratively convenient multiple-choice testing programmes set in. Many current testing proposals include provisions for 'authentic' assessment techniques, also called performance assessment or alternative assessment, that require students to construct answers, perform, or produce something for evaluation. In fact, more than half of the states are currently planning some type of state-wide alternative assessment (Maeroff, 1991). Authentic assessments are preferable to traditional standardized tests, it is claimed, because they are worth teaching to, do not generate the negative test preparation effects of multiple-choice tests, will motivate the unmotivated (Madaus, 1993a), and focus learning on 'higher order or complex thinking skills' (NCEST 1992: 28). Such claims need to be carefully evaluated. It is not the form of the test that is important in determining the impact of a testing programme on students, teachers, and schools. Rather, it is the use to which test results are put (Messick, 1989).

The most current US educational reform proposal, *Goals 2000: Educate America*, continues the tradition of using test results as instruments of educational reform. Goals 2000 proposes a 'voluntary' system of national, 'world class' standards to be in full use by the year 2000. Key components of the Clinton Goals 2000 programme include:

- All children will start school ready to learn.
- Students in Grades 4, 8, and 12 will demonstrate competency, through test performance, in English, math, science, foreign languages, the arts, history, and geography.
- US students will be first in the world in mathematics and science achievement by the year 2000.

Goals 2000 would be implemented through a proposed system of governing bodies, including a National Education Goals Panel (NEGP), the National Education Standards and Improvement Council (NESIC) and the National Skill Standards Board (NSSB). Although the Goals 2000 proposal presents a compelling vision of national standards, it is important to remember that a test based on 'world class' standards is still a test.

Several issues must be closely examined before we undertake a new national testing programme. First, before any proposed examination system can be considered 'equitable', we must ensure that all students have the resources to meet national assessment standards. Perhaps national delivery standards for social, health, family, and educational support systems should be developed in tandem with national test proposals (Madaus, 1993b).

Second, although performance-based tests may be preferable to traditional standardized tests in some aspects, they are still subject to many technical problems when used in high-stakes situations. Individual-level assessment demands high reliability, something that current performance assessments have not been able to achieve. Vermont's initial experience with portfolios revealed that it was very difficult to maintain consistent rating (Koretz *et al.*, 1991). Additionally, evidence suggests that student performance may be task-specific; performance generalizes poorly across different tasks purporting to measure the same domain (Madaus and Tan, 1993; Koretz *et al.*, 1992). Good authentic assessments are also very costly to develop. Their complex design makes them difficult and time consuming to administer, therefore, fewer domains can be tested.

Finally, the purpose of any new assessment system must be clearly defined. A problem of past testing programmes has been that a single instrument is used for a multitude of purposes. However, different uses require different procedures and different techniques. There are two possible purposes that a new system of exams might satisfy: they can be used to provide policy makers with national, state, district, or school-level accountability information; or they can provide information for high-stakes exit, certification, or entrance decisions about students (Madaus, 1993b).

If a system of exams that provides information about individual students is implemented, such a programme should be phased in slowly and subject to careful, independent, monitoring (Madaus, 1993b). The NCTPP recommended in 1990 the 'development of additional institutional means to examine the quality of tests and assessment instruments and to provide oversight of test use' (NCTPP, 1990). The United States has not yet had institutional means to monitor testing.

The need for an independent monitoring agency that could independently evaluate a testing programme before it is adopted, monitor the programme during use, and consider impact after implementation is urgent. A national examination programme carries with it enormous potential benefits or hazards; we should not proceed with such a system until we put into place an independent mechanism to monitor the consequences. Madaus *et al.* (1992) have described in detail how such an organization might be organized, how it might operate, and what activities it might carry out. Such proposals for an independent monitoring body should be strongly considered by those policy-makers who advocate national tests.

Assessment plays many important roles in US education. What we must now do is properly evaluate, prioritize, and monitor each of those roles. Linking the results of standardized assessments (performance-based or not) to high-stakes decisions about students has the potential to corrupt education if we do not proceed with care.

Note

Material for this paper is drawn, in part, from several previous works by one of the authors. See Madaus (1993a, b); Madaus and Tan (1993); and Haney *et al.* (1993).

1. The huge range in these low and high estimates is caused by the inclusion of the cost of student and teacher time devoted to test preparation in the high

estimate. Because several recent surveys show that teachers do spend considerable time in test preparation activities, such as directly teaching the test objectives, test-taking skills, and even specific test items resembling those on the test (e.g., Shepard, 1990; Romberg et al., 1989; Cannell, 1989; Haas *et al.*, 1990; Smith, 1991) the cost of time spent in preparation activities may be appropriate.

References

Beaton, A. E. and Zwick, R. (1992). *Overview of the National Assessment of Educational Progress.* Paper prepared for the National Center for Education Statistics, Office of Educational Research and Improvement.

R. R. Bowker Company (1970–1991) *The Bowker Annual of Library and Book Trade Information.* New York: R. R. Bowker Company.

Boyer, E. (1983). *High school: A report on Secondary Education in America.* New York: Harper and Row.

Buros, O. K. (Ed.) (1978). *The Eighth Mental Measurement Yearbook.* Highland Park, NJ: Gryphon Press.

Cannell, J. J. (1989). *The 'Lake Wobegon' report: How public educators cheat on standardised achievement tests.* Albuquerque, NM: Friends for Education.

Coleman, J. S. *et al.* (1966). *Equality of Educational Opportunity.* Washington, DC: Office of Education.

Greenbaum, W., Garet, M. S. and Solomon, E. R. (1977). *Measuring Educational Progress: A Study Of The National Assessment.* New York: McGraw-Hill.

Haas, N. S., Haladyna, T. M. and Nolen, S. B. (1990). *Standardised achievement testing: War stories from the trenches.* Paper presented at the Annual Meeting of the National Council on Measurement in Education, Boston, MA.

Hall, E. T. (1977). *Beyond culture.* Garden City: Anchor Books.

Haney, W. and Madaus, G. F. (1986). *Effects of standardised testing and the future of the National Assessment of Educational Progress.* Chestnut Hill, MA: Center for the Study of Testing, Evaluation, and Educational Policy.

Haney, W. and Madaus, G. F. (1992). Cautions on the future of NAEP: Arguments against using NAEP tests and data reporting below the state level. In *Assessing Student Achievement in the States: Background Papers.* Stanford, CA: National Academy of Education.

Haney, W. M., Madaus, G. F. and Lyons, R. (1993). *The Fractured Marketplace for Standardised Testing.* Boston: Kluwer Academic Publishers.

Kreitzer, A. and Madaus, G. (1993). *The Test-Driven Curriculum.* Paper prepared for the NASSP under a grant from the Ford Foundation.

Koretz, D. M., Linn, R. M., Dunbar, S. B. and Shepard, L. A. (1991). *The Effects Of High Stakes Testing On Achievement: Preliminary Findings About Generalization Across Tests.* Paper presented at the Annual Meeting of the American Educational Research Association, Chicago, IL.

Koretz, D. M., Madaus, G. F., Haertel, E. and Beaton, A. E. (1992). *Statement before the Subcommittee on Elementary, Secondary and Vocational Education, February 19, 1992.* RAND, Boston College and Stanford University.

Madaus, G. F. (1985). Test scores as administrative mechanisms in educational policy. *Phi Delta Kappan*, **66**(9), 611–17.

Madaus, G. F. (1993a). A national testing system: Manna from above? A historical/technological perspective. *Journal of Assessment in Education*, **1**(1), 9–26.

Madaus, G. F. (1993b). *A technological and historical consideration of equity issues associated with proposals to change the nation's testing policy*. Paper prepared for the Symposium on Equity and Educational Testing and Assessment, Washington, DC, March 1993.

Madaus, G. F. and Kellaghan, T. (1992). Curriculum, evaluation and assessment. In Jackson, P. W. (ed.) *Handbook of research on curriculum*, Washington, DC: American Educational Research Association.

Madaus, G. F., Haney, W., Newton, K. B. and Kreitzer, A. (1992). *A Proposal for a Monitoring Body for Tests Used in Public Policy*. Paper presented at the Conference on the Evaluation of Test-based Educational Reforms, sponsored by the Center for the Study of Testing, Evaluation, and Educational Policy, Boston College and RAND Institute on Education and Training.

Madaus, G. F. and Tan, G. A. (1993). The growth of assessment. In Cawelti, G. (ed.) *Challenges and Achievements of American Education: The 1993 Yearbook of the Association for Supervision and Curriculum Development*. Alexandria, VA: ASCD.

Maeroff, G. I. (1991). Assessing alternative assessment, *Phi Delta Kappan*, **73**(4), 273–81.

Massachusetts Historical Society Documents (1845–46). Horace Mann Papers, Microfilm collection 372, Reel 8.

Messick, S. A. (1989). Validity. In Linn, R. L. (ed.), *Educational Measurement* (3rd edition). New York: Macmillan.

National Assessment Governing Board (1990). Positions of the National Assessment Governing Board on the 1994–1996 NAEP. Atlanta: NAEP.

National Commission on Excellence in Education (1983). *A Nation at Risk*. Washington, DC: United States Government Printing Office.

National Commission on Testing and Public Policy (1990). *From Gatekeeper to Gateway: Transforming Testing in America*. Chestnut Hill, MA: NCTPP.

National Council on Education Standards and Testing (1992). *Raising standards for American education: A report to Congress, the Secretary of Education, the National Education Goals Panel and the American people*. Washington, DC: NCEST.

Office of Technology Assessment (1987). *State Educational Testing Practices: Background Paper*. Washington, DC: Office of Technology Assessment, Science, Education and Transportation Program.

Popham, W. J. (1987). The merits of measurement-driven instruction. *Phi Delta Kappan*, **68**, 680–82.

Romberg, T. A., Zarinna, E. A. and Williams, S. R. (1989). *The Influence of Mandated Testing on Mathematics Instruction: Grade 8 Teachers' Perceptions*. National Center for Research in Mathematical Science Education, University of Wisconsin, Madison.

Shepard, L. A. (1990). Inflated test score gains: Is the problem old norms or teaching to the test? *Educational Measurement: Issues and Practice*, **9**(3), 15–22.

Smith, M. L. (1991). Put to the test: the effects of external testing on teachers. *Educational Researcher*, **20**(5), 8–11.

Snyder, T. D. (1991). *Digest of Educational Statistics, 1990*. Washington, DC: US Department of Education, Office of Educational Research and Improvement.

Tyack, D. and Hansot, E. (1982). *Managers of Virtue: Public School Leadership in America*. New York: Basic Books.

US Department of Education (1991). *America 2000: An Education Strategy: Sourcebook*. Washington, DC: US Department of Education.

West, M. M. and Viator, K. A. (1992). *The Influence of Testing on Teaching Math and Science in Grades 4–12: Appendix D: Testing and Teaching in Six Urban Sites*. Chestnut Hill, MA: Center for the Study of Testing, Evaluation, and Educational Policy, Boston College.

White. E. E. (1888). Examinations and promotions. *Education*, **8**, 519–22.

Wise, A. E. (1979). *Legislated Learning: The Bureaucratization of the American Classroom*. Berkeley, CA: University of California Press.

CHAPTER SEVEN

Inclusive national testing: Chile's 'quality of education assessment system'

Josefina Olivares[1]

This chapter describes a major innovation in Chilean education: SIMCE (Sistema de Medicion de Calidad de la Educacion), or the 'Quality of Education Assessment System'. It outlines the structures and stages of the assessment process, as well as the delivery of results and subsequent actions. Successes, weaknesses and strengths of the assessment process, especially in its contribution to training and monitoring, are highlighted.

SIMCE covers both the basic and intermediate levels of schooling in Chile. However, this chapter concentrates on the first or basic level which has existed since 1988, rather than the second which was only started in 1992. The process and the results that are presented are for 1992 and later, 1992 being the year in which the Ministry of Education of Chile undertook full responsibility for the system.

The Chilean educational system

The Chilean educational system has been organized into four levels: nursery, basic, intermediate, and higher. Basic general education of eight years is compulsory. Intermediate is offered in two forms: humanist–scientific and technical–professional.

As well as the programme for normal education, there are also programmes for out of school, compensatory and adult education. The structure for the system is shown in Fig. 7.1.

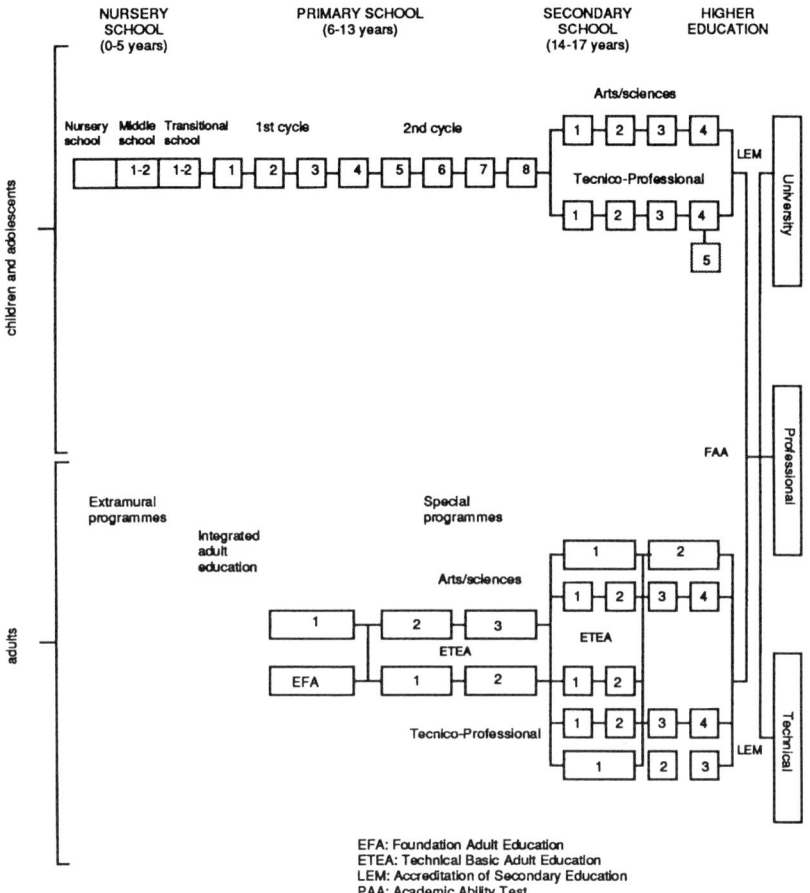

Figure 7.1   The Chilean educational system.

Creation of a system for measuring the quality of education

SIMCE is the logical outcome of previous attempts at evaluation of Chilean education. Towards the end of the 1960s, it was considered necessary to assess the effectiveness of basic education, and to this end, all pupils in the eighth year of basic education were tested on verbal and mathematical ability. However the results of this process, called the National Test (Prueba Nacional), did not have any influence on the conduct or organization of schooling.

In 1982 the Program for the Evaluation of School Results (PER) was outlined and carried out. The lack of awareness of the importance of this programme meant that it only lasted two years. Another attempt, SECE, was carried out in the years 1985–86, under the auspices of MINEDUC (the Ministry of Education). However, in spite of the fact that sensitivity to the problem had apparently increased, the lack of an infrastructure of financial resources along with other factors led to the failure of this campaign.

Bearing past experiences in mind, it was decided to start up the SIMCE through a formal agreement between the MINEDUC and the Pontificia Universidad Católica de Chile. In this way, SIMCE was created in which the university would have most of the responsibility during the first three years. During the fourth year the responsibility would be shared with the team from the MINEDUC, and then it would be transferred to the MINEDUC team through the university's consultant body. Since 1992 the SIMCE has been totally controlled by the MINEDUC.

SIMCE's objectives

Chile's SIMCE is—with government approval—an evaluation programme designed to guide decision making for the different levels of the school system. Therefore, it should provide valid, reliable and suitable information to those who are involved in the educational process.

The established objectives for the SIMCE are as follows.

At the central level of the MINEDUC

- To aid the task of guiding and standardizing the educational system.
- To direct technical and economic support towards the more needy sectors with special attention to the aims of the Program for Improving Quality and Fairness in Education (MECE).
- To contribute information to the Experimentation Center of Pedagogical Improvement and Research of the MINEDUC so as to support its relevant objectives.

At regional, provincial, and community levels

- To aid and guide the task of supervision, technical and economic support throughout educational establishments so as to improve quality and fairness.

At institutional level

- To provide information with which to analyse needs and problems so as to carry out a plan of remedial action which would concentrate on the educational plans of the establishment.

- To involve head teachers, teachers, parents and representatives in the analysis of results and remedial action.

Moreover, SIMCE has at its disposal a data-bank/base open to all institutions and those who are formally authorized to carry out research/investigations in the widest sense of the word: universities, private institutions, members of parliament, political parties.

Functional organization

In obtaining national data, SIMCE depends directly on the subsecretariat of MINEDUC. Its organization is as follows.

Central team

A national coordinator is the executive head of the team. He or she ensures that the objectives are met and organizes the evaluation of the system's conceptual and methodological paradigms so as to promote a valid SIMCE where the country's requirements are concerned.

There are three specific coordination teams who carry out the relevant tasks in an interdisciplinary way:

- *Technical coordination* which arranges all the work needed to prepare the assessment instruments. It has executive powers over five evaluation committees: mathematics, Spanish language, science, history and geography, and personal development. It coordinates the task of the survey, and the analysis and interpreting of the results. It produces the contents of documents, posters, and technical leaflets.
- *Logistical/financial coordination* responsible for all the costs and administration involved in applying the instruments and preparation of reports. It coordinates distribution.
- *Information technology coordination* that is responsible for the whole computer process of the survey, optical reading, programming, the provision of results and reports.

There is also a statistics consultant, as well as an administrative pedagogical consultant. A research committee is in charge of suggesting scientific investigative work, and of linking SIMCE to other research institutions. A technical secretary completes the team.

The skeletal team

The 40 provinces of Chile are grouped into 13 regions. Each region has a regional education secretary; each province, a provincial director. In each secretariat and in each provincial department, there is a civil servant (generally the pedagogical supervisor) who takes on the role of SIMCE coordinator. These 53 coordinators are the executors, at the core of the system, for the stages of the survey, the application of the exercises and the promotion of remedial action to be taken in response to the results.

Costs

Central SIMCE takes advantage of MINEDUC's infrastructure, occupying 630 square metres of it, of which 312 square metres belong to the distribution headquarters. It also makes use of part of the pre-existing furniture of MINEDUC. The cost of the SIMCE-92 application at the basic education level (234,241 pupils grouped in 8062 year groups in a total of 4902 establishments) was $1,009,886 ($1 = $365 Chilean, 1992), on staffing and materials and consumer services. The national experimental SIMCE-92 application at the intermediate education level (48,827 pupils grouped in 1624 year groups in 1624 establishments) was $338,742 for the same items as for basic. Approximately 35% of these costs are accounted for by a grant from the World Bank through the MECE programme, which will eventually be able to move into force in 1996.

Areas of assessment

As noted above, SIMCE has applied to basic education since 1988, and to intermediate education experimentally in 1992, and permanently thereafter. For technical and pedagogical reasons assessment is carried out in the fourth and eighth years of basic education, and in the second year of intermediate. The average age of students in these classes is 9, 13 and 15, respectively.

The assessments relate to different indicators: academic attainment, personal development, attitudes to educational work, and school efficiency. Where academic attainment is concerned, the curricular objectives of the given level or year are considered along with a small proportion of the objectives included in previous (lower) levels. This allows us to have information regarding the educational unit and to contribute to the direction of preparation and remedial action to be developed by the educational centre.

Tables 7.1 and 7.2 give summary statistics for SIMCE in relation to the overall school population.

The students are drawn from all parts of the Chilean education system. For example, in 1992, 58.6% of the 4th year Basic students assessed came from state schools, 33.8% from subsidized private and 7.6% from fee-paying private schools.

Table 7.1 Summary statistics for SIMCE

	Total school population in basic education assessed		Fourth-year pupils assessed		Eighth-year pupils assessed		Total school population enrolled in intermediate education		Second-year intermediate pupils assessed		Second-year intermediate pupils	
	1992		1992	1994	1994		1992		1992		1994	
	2,034,839		234,241	224,678	235,731		675,073		(Pilot)		82,114	

Source: Planning and Budget Office, MINEDUC; SIMCE, MINEDUC.

Table 7.2 Summary statistics for SIMCE

1992		1992	1994		1994
Fourth basic		Second intermediate	Fourth basic		Second intermediate
No. schools	No. year groups (classes)	No. schools	No. year groups (classes)	No. schools	No. schools
4902	8062	1624	7659	4584	1634

Source: Planning and Budget Office, MINEDUC; SIMCE, MINEDUC.

The survey and quality indicators

In addition to individual-level data from students, information is collected relating to the students' schools and areas. A specially designed record sheet allows the collection of information from establishments as follows:

- identification
- characteristics
- geographic catchment area
- administrative organization
- socioeconomic status of the parent body.

Under the heading 'administrative organization', the variable 'dependence' is included, which divides establishments into state, subsidized private and paying private, according to the classification that MINEDUC uses.

The information gathered allows the tests to be grouped by structures. There are 16 functional structures. This allows the organization of information and permits comparative analyses between institutions which are basically similar in terms of population provided for, geographical situation, and financial resources available.

The indicators chosen by SIMCE as suitable for measuring the quality of *basic education* are as follows.

Educational attainment

Educational attainment is characterized by the levels that the pupils reach with respect to curriculum objectives selected as fundamental within the context of the objectives made explicit in the ministry's official plans and programmes. Subjects included are Spanish, mathematics, science and history and geography. There is an assessment instrument for each subject, plus an additional composition instrument for Spanish.

Personal development

This refers to the concept the pupils have of themselves and their surroundings, and is related to the affective climate in which the child is based.

Acceptance of educational work

This aims to discover how the educational unit perceives the administration that runs it. This is bound up with the particular educational aims of each establishment and/or with the specific characteristics of the school community. Different instruments collect the differentiated answers of the pupils, the teachers, the parents, and the governors.

School efficiency

This is related to the capacity of the educational system to retain and advance the pupil in relation to an 'efficiency ideal'. The efficiency ideal is that children should attend and pass eight years of schooling *without grade repetition*.

For *intermediate education*, the indicators are the same as these but measures of 'learning strategies' are added, aimed at establishing how pupils study. The purpose is to direct action and improve the management of learning strategies. In 1993, an ecological indicator was added called 'Attitudes towards the Environment'.

SIMCE produce the assessment instruments themselves, and use test methodology to ensure their scientific rigour. Reliability of the instruments is generally around 0.90 (Kuder-Richardson 20).

As noted above, attainment is measured in four disciplines: Spanish, mathematics, science, and history and geography. Spanish includes a composition exercise. An 'evaluation committee' for each subject, shaped by SIMCE and university evaluators, selects the items for a particular test in conformity with a technical plan and contributes to their theoretical value. The result of this process is an 'experimental test' with at least two equivalent forms which is given to a sample of students representative of the level immediately higher than those who will take the final or 'definitive test'. The provincial SIMCE coordinators and the previously qualified examiners administer the different forms of experimental test under the general supervision of SIMCE staff. The evaluation committees of the four subjects study the answers, the statistical behaviour of each item, its internal discrimination, degree of difficulty, percentage of non-response and final adjustments. The composition test takes a rather different form and can be characterized as a response development instrument.

In the case of personal development, the relevant committee again produces items, perfects those which already exist and sets up an experimental questionnaire to be duly validated. For learning strategies, the instrument of Vladimir Rojas Ossio and others from the Catholic University is used. This professional is a member of SIMCE.

For acceptance of educational work the same personal development committee prepares and validates the three opinion questionnaires which have the following points in common: identification with the school; communication and participation; teacher–pupil relationship; and aspects of the curriculum. The questionnaires are aimed at the pupils, the parents, governors and teachers, respectively. Finally, for school efficiency the planning and budget department collects the data relevant to this assessment through a carefully devised form sent to all the establishments in the range measured.

The instruments are printed (bearing in mind security measures) by private companies who acquire the work through public bidding. In 1992 a total of 2,251,729 instruments were printed.

Raising public awareness

There is diminishing public resistance towards assessments in education. 'Why should we measure the quality of education if we already know that it is very bad?' is an expression which summed up the attitude of the public until a few years ago. Today, things are changing. However, SIMCE remains very active in its efforts to overcome resistance, to make the different audiences sensitive to the value of assessment and to motivate and involve them through their understanding of why measuring the quality of education has been regarded as a *sine qua non* for the future of Chilean education, and the goal of **improving quality and fairness in education with everyone's participation.**

What has been done? SIMCE does not have a social communication and diffusion substructure; therefore the staff of the SIMCE sensitizes, motivates, involves and invites participation, making the most of all opportunities as they arise. Nevertheless, there are allies that SIMCE did and did not intend to have. The highest authorities of MINEDUC, including the minister, have declared their intention of using assessment to improve the quality and fairness of education. Through national and regional press conferences they have informed the country of forthcoming national applications of the SIMCE test and its aims. In the same way, a substantive part of the results obtained are communicated to the country in general, and particularly to the regions.

The press has responded positively to these communiqués. A salient example is that in 1991 it was the minister himself who asked SIMCE to speed up the work on assessment in intermediate education. The political class of the country—reappearing after a prolonged authoritarian regime—and comprising members of parliament, mayors, councillors and political party members, ask for and use the SIMCE results to gain credibility. Central, regional and provincial MINEDUC staff use the results to help them to monitor the evolution of the relevant educational aspects and to react with resources to the urge for improving standards and fairness.

Every year a poster appears in offices, schools, and some public places across the country. For example, in 1992, 10,000 posters were distributed, along with 174,000 technical leaflets for executives and teaching staff, 12,300 for examiners and 13,000 pedagogical guidance manuals for teachers. To this is added the distribution of 11,000 three-part documents for executives and teaching staff; and 300,00 three-part documents for parents and guardians.

Every year there is a meeting of provincial and regional SIMCE co-ordinators with the central team to evaluate the task demanded by the assessments, the results and remedial action. Central SIMCE receives feedback. Both in the application of the test and in the submission of results, the motivation of different members of the educational community—authorities, politicians, teachers' unions—can be different. The results evoke a range of reactions which are positive and negative, destructive and constructive, pessimistic and optimistic. In general, sensitization, understanding and even involvement are increasingly arousing more expressions of 'assessing so as to improve'.

The least sensitized or involved sectors in this resounding binomial 'assessment—quality improvement' relationship are the poor suburban or rural ones. However, there is no doubt that overall teachers are surmounting the rejection of assessment centred on pupils.

Implementation

Continental Chile is 4270 kilometres in length and, on average, 180 kilometres wide. The distribution headquarters of SIMCE mobilizes across about 11,000 kilometres of paved road network, and almost 70,000 kilometres of crumbling roads and dirt tracks. Over 2.25 million instruments, reply sheets, and control sheets are distributed throughout the area for basic education; and for intermediate education, over half a million of the same type. Still to be added are the hundreds of thousands of leaflets, three-part documents, and posters. The headquarters has a modest infrastructure of 312 square metres. The distribution of instruments and their immediate return after completion are carried out over a maximum of 30 days using the private company that bid for the contract, but supervision is carried out by SIMCE. Security measures have been taken so as to protect the confidentiality of the instruments on the outward and return journeys.

The highest degrees of secrecy need to be maintained during production, distribution and destruction of the instruments so as to be able to obtain maximum confidentiality. The security factor is very difficult to achieve to a high level of purity. Every year, new weak points are discovered. The SIMCE professionals' level of awareness of the importance of confidentiality is high. It is less high among others who are part of the process, for example the companies who print instruments, process the data, print reports, distribute instruments throughout the country, etc. Members of the SIMCE team are involved heavily in:

- custody of the items at every stage of its production
- custody of experimental and definitive instruments at every stage of their production
- suitable security in private companies awarded the contracts for printing the instruments, processing the data, printing the reports with results
- strict supervision during all the stages of applying the instruments
- custody and security system in the instrument storage and distribution headquarters.

Administration is a large-scale exercise which rests ultimately on the quality of SIMCE staff. For example, in 1992 the administration of the basic education tests involved 8062 examiners selected according to a specific profile, and trained through two sessions by 40 provincial SIMCE coordinators. Armed with a manual with all the instructions they attend 8062 classrooms, already previously visited and organized by them.

Scheme supervisors from the provincial departments cover the area. In the same way, all of SIMCE's technical staff carry out the task of supervision, leaving only the technical coordinator at SIMCE's headquarters to deal with emergencies. After the rigorous counting of the instruments, answer sheets and within a maximum of 30 days, the application material is returned to the distribution headquarters of the SIMCE. In general, pupils' receptiveness to the tests and questionnaires has been excellent. In 1992, out of 256,697 children surveyed, 234,241 replied to the basic SIMCE application.

Results and reports

The increasing degree of awareness among different users means that the results of the assessment are awaited and even demanded. SIMCE reads the answer sheets and the classification of the composition instrument optically. The processing of data validated by SIMCE is carried out, with varying degrees of delay, by a private company the contract through public bidding. This reduces the speed with which SIMCE can submit the results.

Another public bidding round awards another private company the printing of the school, provincial and municipal departments' reports.

While this is happening, reports are passed on to the highest authorities of MINEDUC which, calling a press conference, distributes the more relevant aspects shown in these results to the nation. The basic reports of these national and regional communiqués refer especially to academic achievement and contain standardized national and regional results for academic objectives, by subjects, and by type of school (private paying, subsidized private, and state establishments).

The computer printed, colour-coded reports which are distributed to the whole country (using private transport, bid for publicly) are as follows:

- *To regional education secretaries*: reports with the results of all the indicators on a regional level and provincial departments that make it up.
- *To the provincial education departments*: reports of all the indicators on a provincial and community level. As well as this, a synopsis of achievement reports in a list of the establishments of the province.
- *To the local authorities*: report with results of all the indicators at a local authority level.
- *To educational establishments*: report with the results of all the measures on a year group and establishment level. Report with the general results of the establishment for parents and guardians. A manual with pedagogical guidance to support remedial action completes the set of reports.

Other than results, the reports contain conceptual clarification and instructions for use.

Tables 7.3a–c provide illustrations of the information available from SIMCE. They show the mathematics scores, by deciles, of the 1993 eighth-year basic education students, broken down by type of school, type of community, and socioeconomic status of the parent body. As noted above, the mathematics assessments (and scores) are constructed in relation to national curriculum objectives.

The results illustrate clearly the gap between, for example, high-level socioeconomic level private schools in large towns, and state-subsidized schools in low-income country areas.

Utilization of the results

SIMCE results are gradually being used in the intended way. On the basis of SIMCE information, an intensive programme to help basic schools in the poorer sectors has been promoted from the central level. Out of a total of 8000 basic education schools in the country, about 900 from rural and poor urban areas were chosen for intervention. During 1991 this programme was planned to raise the school performance of 1385 basic schools

Table 7.3 Eighth-year measurement of gross scores in deciles by socioeconomic status, type of city, and school status. (Maths scores)

Socioeconomic status	Type of city	No. of schools	1	2	3	4	5	6	7	8	9	10
(a) Private schools												
High	Capital	132	49.70	60.30	67.10	71.86	74.50	77.30	80.13	82.27	85.20	90.50
	Large	102	54.70	64.20	69.70	72.66	75.17	78.34	81.70	83.42	86.26	91.50
Medium	Capital	66	46.30	56.14	60.30	64.90	68.00	71.10	75.60	79.70	82.30	90.50
	Large	103	52.15	58.82	63.95	66.14	70.33	72.47	75.78	78.35	81.67	85.50
	Others	1*										
(b) Private with state subvention												
High	Capital	13	51.80	53.10	53.80	54.23	58.00	60.30	62.55	63.20	73.20	85.50
Medium	Capital	165	44.00	49.70	51.46	54.06	56.57	59.83	63.67	67.38	73.67	86.50
	Large	257	46.70	51.56	54.72	58.04	60.92	64.52	66.49	70.22	75.88	84.50
	Others	64	46.90	50.40	54.14	55.80	58.30	61.10	64.40	65.60	69.37	75.50
Low	Capital	279	40.85	43.38	46.14	48.46	50.18	52.10	55.03	58.19	64.05	85.50
	Large	234	41,60	44.92	47.63	50.68	53.25	56.26	58.41	61.54	66.43	78.50
	Others	216	37.02	39.86	43.92	46.88	48.21	50.77	53.53	57.00	60.90	76.50
(c) Public												
High	Capital	1*										
	Large	6*										
Medium	Capital	74	42.70	46.10	48.59	51.30	54.10	56.10	61.30	63.10	68.10	86.50
	Large	192	47.03	50.13	51.87	54.28	57.38	59.60	61.78	64.86	70.43	81.50
	Others	57	46.20	49.10	51.52	52.77	54.60	55.80	58.13	61.10	62.38	70.50
Low	Capital	354	40.66	42.99	44.79	46.79	48.42	50.33	52.72	55.22	59.30	73.50
	Large	727	42.48	44.94	47.10	48.84	50.64	52.41	54.67	57.25	60.69	78.50
	Others	1276	39.58	42.96	45.31	47.51	49.48	51.51	53.69	56.50	68.98	86.50

*Deciles in cases with less than ten schools have not been calculated.

(15.1% of the total) benefiting 222,491 children and supporting 7267 teachers (Statistical Information Compendium 1991, MINEDUC, Chile).

At provincial and community levels, the SIMCE results provide information for decisions about technical and economic support and the supervision directed at the educational process of each establishment.

The 'school classroom' is the gravitational centre for the productive use of the results. SIMCE provincial coordinators try to ensure that more than 80% of the establishments which receive results discusses them at least at teacher level. Approximately 60% reach conclusions and plan some form of remedial action. Around 40% carry out what can really be called remedial action and about 30% carry it out in an organized manner according to a plan. A comparison of the SIMCE scores from 1988 to 1992 show, especially in Spanish and mathematics, that there are differences reflecting improvement. Members of Parliament, mayors and political parties use the SIMCE information to different ends which, democratically, should contribute to the improvement of the quality and fairness of our educational system.

The results are analysed and interpreted by the SIMCE team, through an increasingly technical and demanding task aimed at improving effectiveness and efficiency of SIMCE itself. It is an ongoing project to increase SIMCE's contribution to the political and educational requirements of the country, for example through the production of a pedagogical guidance manual for teachers, methodological, logistical and financial corrections of the assessment process itself.

SIMCE does not carry out any pure research into its data, but, being the property of MINEDUC, it is available to any person or institution aptly authorized to carry out relevant research. The MINEDUC Center of Pedagogical Improvement, Experimentation and Research has begun to conduct such research. An *ad hoc* committee outlines possible research to be suggested to bodies especially universities.

Current issues

The assessment system in Chile should not stagnate and needs to evolve within current levels of resourcing. The demand for new indicators (for example the ecological indicator 'Attitude towards the Environment') and the need to measure other levels of the system and initiate relevant research, mean that SIMCE should reconsider the intervals at which assessment takes place, and will need to give preference to the sample format within the framework of submitting results by establishment. In future, it should alternate measures but always within the framework of its intentions and decided objectives and create time for a minimum amount of research.

Immediate issues of concern include the following.

'Fossilization' of the curriculum

This was how, after his visit to SIMCE, Protase Woodford summed up the fact that the content of the instruments is based on objectives derived from the current curriculum which may not be the most appropriate for Chile. There is the risk of a vicious circle in which the instrument would measure, with a high degree of accuracy, counterproductive objectives (report to MINEDUC). SIMCE agrees with this authorized opinion. Fortunately, MINEDUC is closing the *consesualizacion* of a proposal phase—coming from a constitutional education organic law—which referred to the 'fundamental objectives and minimum contents of general basic and intermediate education'. In this document not only are some academic objectives renewed, but also objectives are included in the area of effective and social personal development. It was in this way that the new SIMCE indicator 'Attitudes Towards the Environment' was guided. With the new rule on fundamental objectives and minimum content being introduced, SIMCE will have to carry out adaptation studies. Work has begun on this.

The bidding process

As explained above, the processing of data, the printing of reports and the distribution of the instruments are all awarded through public bidding to private companies. Usually these companies either pull out or their work is deficient. This creates stress for SIMCE, since it is their responsibility to carry out a dependable and tight work schedule which is essential for the efficiency of the assessment.

Turnover of professionals

This problem afflicts computer engineers particularly. Given that private companies pay substantively better than MINEDUC, the engineers contracted by SIMCE—who are themselves paid better than the rest of the staff—receive the training, learn, and then migrate to the private company. In this way, there is a great source of tension which affects the flow of tasks in the other technical and logistical areas of SIMCE. As yet, there is no solution to this problem, which is similar to that experienced by other public sector projects (in Chile and worldwide).

Logistical fatigue

The enormous scale of SIMCE is another variable which prompts us to favour a sample rather than population method in the future.

Negative impacts of the SIMCE results

Until now these have not been so significant that they threaten the stability of the assessment system. There is still time to achieve a better culture of evaluation among teachers. There are teachers who develop their teaching programmes as if they are 'preparing for the SIMCE' and there are even cases in which they have tried to 'help' their pupils in unorthodox ways. More serious still is the fact that there are establishments where teachers have been asked to leave when the results do not match up to optimistic hopes. In the same way, there are heads of establishments who do not pass on their poor results to governors in case the latter could create difficulties. Finally there are the different sections of the social and school communities who tend to make a 'league table of good and bad schools'. Having said this, SIMCE does not skimp on effort to increase positive awareness of the reasons behind the assessment.

Deficient use of SIMCE data in research

The most competent body for this should be the MINEDUC Pedagogical Improvement, Experimentation and Research Center, to which all data are sent every year. This centre and SIMCE have started up a research programme which will be carried out by the centre and which SIMCE will support. Universities hardly use the data bank. SIMCE has created a research committee in charge of its promotion.

Importance of addressing technical and vocational education

There is a wide consensus that technical and vocational education at the intermediate level should be paid special attention in the near future in the country. SIMCE is aware of the need for its participation and how to take on this difficult and complex sector is beginning to be high on the agenda within the context of its development.

Note

1. Translated from the original Spanish by Marta Monzon.

CHAPTER EIGHT

Assessment developments in England and Wales: the triumph of tradition

Patricia Broadfoot and Caroline Gipps

Since the introduction of mass schooling in the latter part of the nineteenth century, educational provision in England and Wales has been characterized by the domination of formal assessment practices. These practices have taken a variety of forms. Her Majesty's Inspectorate (HMI) of Schools was founded in the early nineteenth century with a remit to judge the quality of educational provision. Later in 1857 the first 'public' examinations were introduced to provide a standardized means of accrediting school-leaver achievements and to provide for university selection. Later still, after 1870, scholarships began to be introduced to open up the possibility of a 'secondary' education to those judged able to benefit from it but unable to pay.

These three forms of assessment—inspection, certification and scholarships together provided a powerful source of control over an unruly education system that was to the outside observer breathtakingly free of central control. With no centrally-prescribed curriculum, no national controls over funding and only the barest minimum of formal requirements for provision enshrined in legislation, Local Education Authorities (LEAs) and schools in England and Wales have both traditionally been relatively free to shape the educational provision in their respective spheres of operation. However such shaping has taken place within the limits of the substantial constraints exercised by formal assessment arrangements.

It is important to understand these defining traditions of educational provision in England and Wales in order to fully appreciate the impact of recent policy changes, especially those relating to assessment practices most of which have as a defining principle the strengthening of central government control. Although the introduction of a national assessment programme as part of the provisions of the 1988 Education Reform Act was particularly explicit in this respect, other recent policy initiatives such as the introduction in 1988 of the new public examination at 16+—the

General Certificate of Secondary Education (GCSE)—have been characterized by similar goals.

In this chapter we first consider briefly the traditional structure of educational provision in England and Wales and then go on to document some of the significant changes that have taken place in recent years in relation to assessment. By so doing we hope to illustrate how the enduring tensions between the various purposes of assessment which are the subject of this book have characterized educational policy decisions in England and Wales.

Figure 8.1 summarizes the education system of England and Wales as it largely still exists. Children enter primary school in the year in which they will become five years old. At the age of 11 years most will progress to a non-selective 'comprehensive' secondary school until they are at least 16— the minimum age for leaving school—with increasing numbers staying on until they are 18 to study for GCE A (Advanced) level or the recently introduced General National Vocational Qualifications (GNVQ)—a kind of vocational A level. At 16, almost all students will take the GCSE examination. A few children are also still required to sit for a selective examination at age 11. This hangover from a previous era in a few LEAs well illustrates the continuing scope for local decisions concerning the nature of educational provision and the diversity of practice to which it can lead.

Whilst the broad structure of the educational provision illustrated in Fig. 8.1 derives from the 1944 Education Act which established a system of universal secondary schooling for England and Wales, the content of the education which is delivered at these various stages has been substantially altered by the major legislative innovations of the 1988 Education Act. This initiative was designed to strengthen central control of both the content and the quality of educational provision.

At the heart of these developments was a concern about educational standards in terms of the range of curriculum experiences offered to pupils in different schools, the rigour of teaching in the basic skills, and low expectations for pupils' performance. Both the first and last of these three had been a regularly-voiced criticism by the then independent HMI in England and Wales. In reality there was less curricular variation at secondary level than at primary level since, as has already been suggested, the upper secondary school curriculum has long been controlled to a great extent by public examinations at 16+ and 18+. The concerns at secondary level were more that pupils were dropping subjects, in order to specialize, as young as 13 or 14 and that the range of curricular provision for the bottom 40% of the ability range was inadequate. In addition, with the advent of comprehensive secondary schools during the 1970s and consequently, the virtual abolition of the 11+ selective examination, there was no 'hard' information on the performance of primary schools.

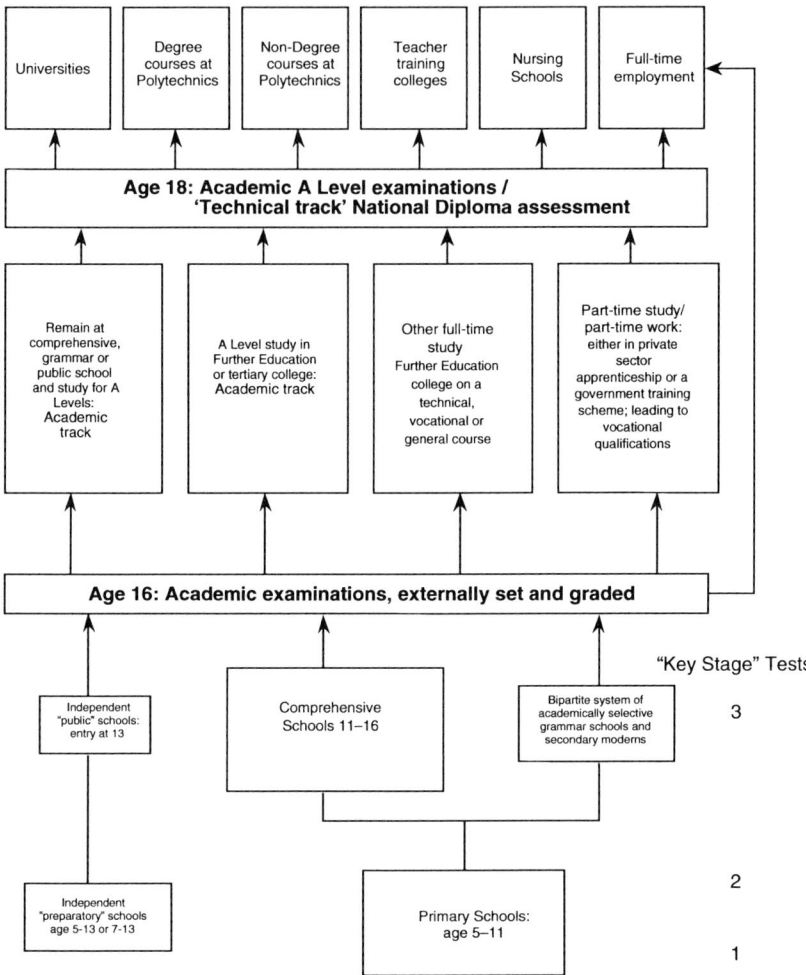

Figure 8.1 Structure of the education and assessment system, England and Wales.

A significant factor in the call for an improvement in educational standards was a report published in 1983 comparing performance in mathematics standards in schools in England and West Germany. The authors reworked data from the 1964 IEA Study (see Chapter 4) and claimed that German pupils in the bottom half of the ability range obtained levels of performance comparable with the average for the whole ability range in England (Prais and Wagner, 1983).

A number of other international comparisons also showed that English schools were not top of the league tables. The previous national assessment programme, the Assessment of Performance Unit, which had carried out anonymous testing of 'light' samples of pupils, during the 1970s and 1980s had been unable to comment satisfactorily (because of measurement problems) on whether national standards were rising or falling. These other studies shifted the argument away from comparisons over time to comparisons of the standards achieved by English schools with those of other countries. Politically this was a more powerful argument within the context of discussions about economic decline.

For the first time in the history of English education therefore it was decided in 1988 that a national curriculum was to be introduced. This was to be designed to ensure that all pupils of compulsory school age (5–16) would follow the same course with English, mathematics, science and technology forming the core, and history, geography, a modern foreign language, art, music and physical education identified as 'foundation' subjects. These ten subjects together were to make up 70% of curriculum time.

The curriculum for each subject was enshrined in law: statutory orders described the 'matters, skills and processes' to be taught as 'programmes of study' and the knowledge, skills and understanding making up 'attainment targets' (ATs) within each subject which pupils were expected to have reached at each stage of their schooling. These stages were defined as key stage one (age 5–7), two (7–11), three (11–14) and four (14–16). The ATs were described in a series of ten hierarchical levels in order to facilitate progression. Thus most pupils of 7+ would be at level two in the system while most pupils of 11+ would be at level four and so on. The ATs were articulated at each of the ten levels by a series of criteria or 'statements of attainment' (SoAs) which formed the basic structure of a criterion-referenced assessment system.

The introduction of a national assessment programme was a crucial accompaniment to the national curriculum, since it was intended to provide for both performance to be measured and standards to be raised. It built on the long-standing policy tradition of using assessment as a means of influencing priorities and practices within the system. In 1927 Selby Bigge argued that 'we must look to examinations rather than inspection to check, test and secure the efficiency of public education' (quoted in Silver, 1979). These sentiments were held even more strongly by the Conservative government of the late 1980s, enmeshed as they were in an international policy climate in which concerns for educational competitiveness on the one hand and accountability on the other had produced an obsession with assessment that was far from being confined to England and Wales. But, if the use of assessment as an explicit policy tool was not, in itself, novel in the context of England and Wales, its form and purpose certainly were. The introduction of formal national assessment at every stage of the education system, including for seven-year-olds in the infant school, was rooted in a desire to generate a national 'currency' which could be used in a quantitative way to identify and compare individual, institutional, local and national standards.

Where other countries were content simply to use such information to *inform* policy decisions, the strong political commitment of the British government to the operation of 'market forces' as a means of raising standards through competition led to national assessment being explicitly designed to encourage competition between schools.

The details of how this was done, why and with what effects, is the subject of the next section of this chapter. It is a story of both hope and despair; of vision and of lost opportunities; of triumph and of disaster; and of political will against professional commitment. Above all it is a story that needs to be understood in its context; as part of the broader setting of professional traditions and policy assumptions which have long characterized education in England and Wales. The advent of national assessment in 1988 represented both a continuity with these traditions and, at the same time, a profound rupture with the past.

The national assessment programme in England and Wales

The national curriculum assessment programme is the largest assessment development to be undertaken in recent years in the UK, requiring as it does detailed assessment across the core subjects of the national curriculum at ages 7, 11, 14 and 16. The purpose of the assessment programme is to measure the performance of pupils against the national curriculum; to provide accountability data for schools and local education authorities so as to raise standards of performance; and to support the teaching–learning process. The original proposals for national assessment involved a combination of teachers' own assessments and external 'performance-based' tests. There has, however, been a considerable retreat from this model over the subsequent years of development. Before giving a detailed account of these developments and of the reasons for the various shifts in emphasis that have taken place, it is necessary to give an outline of the evolution of the assessment model and how it relates to the curriculum structure.

The first stage of development was the setting up of the Task Group on Assessment and Testing (TGAT) with a remit to design the assessment programme. The report of this group (DES, 1988) put forward a blueprint to which all subjects had to adhere. The national assessment programme, as outlined in the TGAT report and the statutory orders, originally required that pupils be assessed against all the ATs (Attainment Targets) by their teachers and on some ATs by external standardized assessment tasks (SATs) at the ages of 7, 11 and 14. At these ages the results of teacher assessment (TA) and the SATs, were to be combined and reported. At 16 the external test was to be the GCSE, the public examination which is taken by approximately 85% of the age group. The grading system of the GCSE was to be merged with the ten level national curriculum scale.

At the individual level, national assessment results had to be reported to parents. This requirement subsequently became incorporated as part of the implementation of the Parents' Charter which required all schools to report annually on all children in relation to every national curriculum subject, including comments on general progress and a record of attendance (Circular 14/92). At the end of each key stage the pupil's performance had to be reported in terms of levels (including at age seven, separate arithmetic, spelling and reading levels) and comparative information had to be given about the achievements of other pupils of the same age/stage in that school. The availability of such comparative information in school prospectuses, for example, which is now a statutory requirement, readily provides for comparisons to be made between schools.

Indeed, since 1992, reporting on school performance (Circular 7/92) has been structured specifically to allow comparative tables of school performance in public examination results at 16 and 18. These are distributed by primary and middle schools to parents of children about to transfer to secondary school and published by the government in local newspapers. A full statement of their public examination results has to be made available by each secondary school at least two weeks before parents' choice of secondary school has to be made. Average figures for the whole of England are supplied to governors to go in school prospectuses. The same procedures will eventually apply for national assessment results at the end of key stages two and three as well.

The only additional feature of the national assessment arrangements for Wales is that Welsh is assessed both as a first language *and* as a second language. SATs are therefore available in both English and Welsh. In Welsh medium schools, pupils are assessed using SATs on mathematics, science and Welsh at seven, and these subjects together with English at 11 and 14.

The first phase of the implementation of national assessment was that for 7-year-olds at the end of key stage one. After a pilot in 1990 the first full run of assessment was in 1991 with key stage three due to start in 1993. In the event however, the opposition that had been steadily mounting as infant teachers struggled to implement procedures which they perceived to be disruptive, time-consuming and unnecessary came to a head with the extension of national assessment to 14-year-olds. Battles over what was deemed to be appropriate content and style for such tests grew increasingly intense. The government's desire for traditional, knowledge-based written tests clashed with the widespread professional belief, especially among English teachers, that the curriculum should emphasize skills and understanding and that the tests, likewise, if they had to exist at all, should represent a valid reinforcement of these priorities.

The more detailed account of the early experience of national assessment at key stage one which follows, well illustrates these tensions and hence the reasons behind the unprecedented decision on the part of the mass of teachers and schools to boycott national assessment in 1993.

During the spring and early summer term of the year in which pupils reached the age of seven (year two) teachers were to make an assessment of each pupil's attainment in relation to the ATs of the core subjects in terms of the ten-level scale. Teachers were to make these assessments in any way they wished, but observation, regular informal assessment and keeping examples of work were all encouraged. In the second half of the spring term and the first half of the summer term the pupils were to be given, by their teacher, a series of SATs covering a sample of the core ATs.

Because of the significant role which was to be played by teacher assessment (TA) in the national assessment programme, the TGAT report had suggested a complex process of group moderation through which teachers' assessments could be brought into line around a common standard and any variation between TA and SAT could be settled professionally. This was deemed to be too expensive however and was never implemented. Instead, visiting 'auditors' were appointed who would visit schools to ensure that both procedures and standards were being applied in a common way. How TA and SAT results were to be combined was from the outset and continued to be a contentious area; they are now reported separately.

Since the proposals for the SATs in the TGAT Report were innovatory and were a conscious attempt to move away from traditional standardized testing procedures they will be described in some detail. The TGAT report suggested that a mixture of instruments including tests, practical tasks and observations should be used in order to maximize validity and hence, minimize curriculum distortion. It was hoped that a broad range of assessment instruments sampling a broad range of ATs would discourage the narrowing tendency to teach to the test. Thus the TGAT model was one which emphasized the use of a wide range of assessment tasks. It provided for a variety of response modes in order to minimize the negative effects normally associated with formal assessment, especially with such young children. It also envisaged a wide range of assessment tasks using different contexts to ensure content and task validity. Thus SATs represented a notable example of the kind of performance-based assessments which are currently receiving considerable attention in the United States and elsewhere.

In the early stages of the development of the SATs for key stage one the requirement was that they should cover as many ATs as possible. This proved unwieldy, since there were 32 ATs in the original curriculum structure for the core subjects alone, especially given that the mode of assessment was to be active rather than paper-and-pencil tests of the traditional standardized type.

In the event, the SATs used with seven-year-olds in 1991 were a watered-down version of the TGAT proposals. The style of assessment was however active and similar to good infant-school practice: for example, the reading task at level two involved reading aloud a short passage from a children's book chosen from a list of popular titles, using dice to play mathematics 'games', using objects to sort and so on.

Despite the reduction in the number of ATs to be tested from 32 to nine, the SAT administration in 1991 took a minimum of 40 hours for a class of 25–30 pupils and was rarely managed without support for the class teacher, since most of the SATs were done with groups of four pupils. Generally, it was felt, that the SATs matched good teaching practice; that they provided teachers with useful and detailed information about individual children, but that they were excessively time-consuming as well as offering limited standardization for comparability purposes (Pollard *et al.*, 1994).

Even before formal evaluations of the SATs were available, in response to the widespread publicity about the amount of time the seven-year-old SATs were taking, the prime minister announced in the summer of 1991 that for 1992 there would be shorter, standardized paper-and-pencil tests. The 1992 SATs contained a reduced number of active tasks, and offered for a number of SATs a 'whole-class' administrative procedure, which in fact few teachers used. The reading SAT stayed as a reading aloud task with the teachers making a running record and in addition an accuracy score. There were also two standardized tests: a traditional group reading comprehension test with a written response, and a group spelling test. The reading test was optional at level two and above and the spelling test was compulsory for level three and above. These two scores had to be reported separately alongside the mathematics 'number' score, as well as the overall levels for English, mathematics and science.

From 1993 spelling and reading comprehension tests became compulsory for all except level one. Different ATs in mathematics and science were to be covered each year in addition to 'number' so that in 1993 seven-year-olds were assessed on algebra and physics. The testing package still took around 30 hours of classroom time as it did in 1992 but the original emphasis on 'performance-type' assessment was being progressively eroded in favour of more traditional paper-and-pencil tests.

However, the evaluation of the 1994 key stage one tasks and tests in English and mathematics nevertheless reported that children typically enjoyed the tests, many still not realizing that they were being tested. Lessons were being learned in schools in the light of experience concerning how the demands of testing could be reduced with cooperation between colleagues. Many schools had experimented successfully with ways of improving comparability including small-scale agreement trials: cross-marking; paired-marking; the sampling of scripts and tests with several classes by the subject coordinator and collaborative marking sessions (SCAA, 1994).

The 1995 arrangements at key stage one now require 'statutory' assessment in English, mathematics and science with national tests and tasks only in English and mathematics. The TA and SAT results are shown separately in all forms of reporting. Whilst the SATs are externally 'audited' by the LEA to ensure consistency of standards and administration, (though not at age seven), TAs have no such statutory requirement. However, informal procedures, such as those described above, are encouraged.

Thus the history of key stage one national assessment shows particularly clearly the tensions inherent in trying to introduce assessment procedures which are both sufficiently rigorous that they can provide a level of reliability that is sufficient for the purposes of comparability and yet valid in terms of curriculum objectives. Add to these technical problems those of teacher workload and ideologically-motivated political interference and it is easy to see why the evolution of national assessment procedures in England and Wales has been so tortuous and difficult. The original TGAT conception of a national assessment programme that could provide for diagnostic, formative, summative and evaluative assessment purposes has been shown to be unworkable as the varying agendas defined for national assessment have proved incompatible in terms of the procedures which are appropriate to each.

National assessment at key stages two and three which has been implemented much more recently, has resolved this particular tension by being clearly intended to emphasize the summative and the evaluative at the expense of the formative and diagnostic. It has reverted to the well-worn path in education in England and Wales of using externally set and marked tests to dictate curriculum priorities and to regulate national standards.

At key stage two, age 11, the emphasis has been, from the outset, on tests rather than tasks. In 1994, English for example, consisted of a reading task for levels one to two using a choice of texts, a reading test for levels three to six based on a magazine, a writing task based on narrative writing for levels one to six and spelling tests covering levels one to two and three to six. As with key stage one in 1992 and 1993, the 'process' ATs in each subject are not covered by the SATs, but assessed by TA. Whilst this may well be a more satisfactory way of assessing these skills, there is a distinct danger of this part of the curriculum becoming downgraded since it is not included in the external 'high stakes' testing.

Similarly, the trialling of SATs for 14-year-olds which took place in 1991 involved extended tasks taking many hours of classroom time and covering a range of activities and response modes. The secretary of state for education deemed this inappropriate and the pilot 'SATs' in 1992 were short written tests done by whole classes at the same time under examination conditions. Practical tests were only to be set where there was no alternative.

As with key stage one the development agencies were first asked to assess each AT through SATs. Not only did this make the test development task enormous, the preferral of SAT results to the TA result where both were available meant that SAT results at the individual pupil level had to be highly reliable—more so than the TGAT report had envisaged (Brown, 1992). Furthermore, the SATs at age 14 had to cover all ten levels of the national curriculum.

For the 1991 trial, the secretary of state for education required an element of written testing taken under controlled conditions. Again, as with key stage one, before the evaluations were complete the secretary of state—who referred to the SATs as 'elaborate nonsense'—announced changes. Contracts with the development agencies were terminated and new contracts were put out to tender. The specifications for 1992 required three one-to-one tests and a half-hour test per subject covering all ATs, except the process ones which were assessed by teachers. Papers were to be set at four levels each covering a range of four national curriculum levels; teachers were to select the level at which to enter a pupil, although they would not see the examination papers beforehand. All the tests were to be taken on the same dates in June in formal examination conditions.

One of the reasons put forward for changing the style of the assessments was the amount of time the original SATs took, the same manageability issue that dogged the key stage one assessments. However, the evaluations showed that the key stage three teachers did not find the task burdensome and felt that the active SATs were a valid way of assessing performance (Brown, 1992; Stobart and Burgess, 1992; Jennings, 1992). It seems clear that the decision to alter the assessments from active, extended perfor-mance-based SATs to timed written examinations was essentially a political one since at these the SATs did not present a manageability problem, were widely felt to be valid and had been demonstrated to be sufficiently reliable. The potential that teachers felt they had to support learning through the provision of diagnostic information on pupils' strengths and weaknesses was sacrificed in favour of the perceived rigour of an externally-set and marked test. In fact there were very obvious technical shortcomings in the marking of the 1995 English tests at key stage three which in some schools were found, on re-examination of the marking, to be 80% wrong. Not surprisingly, many teachers regarded the information so generated concerning pupils' achievements in English to be completely useless if not dangerously misleading.

The early trials of national assessment for key stages two and three were disrupted by the widespread refusal by teachers to conduct them in 1993. So bad had relations become between teachers and the government that the former were prepared to risk legal sanctions in making a stand against what they saw as unreasonable and unprofessional demands by the secretary of state for education.

The result was the launch of a consultation exercise, chaired by Sir Ron Dearing which was to review the scope and operation of the national curriculum and assessment arrangements in the light of the widespread complaints coming from schools that the amount of subject-content was excessive and the demands on teachers of both national curriclum implementation and national assessment were unacceptable in terms of workload. The result, for key stages one, two and three was a national curriculum slimmed to free up 20% of time and up to 40% in key stage four. The time allocated to each subject group was designed to ensure that the whole curriculum could be taught within the total time allocations for each key stage. The number of ATs and statement of attainment (SoAs) in each subject was reduced and the ten-level scale redesigned to end at the end of key stage three. SoAs were to be replaced by more synoptic 'level descriptors' for each subject at each level.

The revised national curriculum was introduced to schools in September 1995 following enthusiastic support for the recommendations arising from Sir Ron Dearing's review. The national assessment arrangements for 1996 were little changed from those of 1995 and confirm the establishment of the external component of national assessment at key stages two and three as being in the traditional mode of external examinations. Although complemented by TA, the absence of a requirement for a formal audit of the standards being applied in the latter is likely to mean that the results of TA—albeit covering key curriculum objectives—are not taken as seriously in 'currency' terms. The perceived security of tried and tested assessment methods coupled with the problems of workload which were initially encountered by teachers in implementing national assessment, have combined to close the window of opportunity for a genuine step forward in assessment practice which the TGAT design opened. Arguably, assessment for measurement purposes has triumphed over assessment for the support of learning itself.

Interestingly, the stories of assessment development at 16+ are very similar and serve to underline the fact that policy decisions concerning assessment procedures are as much, if not more, a reflection of deep-rooted national assumptions concerning assessment purposes and practices—both historical and contemporary—as they are of developments in assessment techniques *per se*.

Assessment at age 16

The GCSE is the public examination currently taken by pupils at 16+. This is itself a relatively new examination for which the first papers were taken in 1988. It brought in many changes including the use of coursework assessment rather than 100% examination so that oral, practical and extended project work play an important part in the assessment. It is aimed at the whole ability range. This means that differentiated exam papers pitched at different levels are required for some subjects. The GCSE was intended to be a criterion-referenced examination so that candidates could be graded in relation to their own performance rather than in relation to how others performed.

It is generally acknowledged that the GCSE has brought about changes in teaching style and content resulting in a broadening of students' curricular and pedagogic experience. A higher proportion of the age group takes it than was the case with the previous 16+ exams (over 85% of the age group enters at least one subject). Coursework assessment has had a powerful effect in many schools and 100% coursework-assessed syllabuses were popular in English and in some other subjects until they were forbidden by the government in 1993. The move towards criterion-referencing has been problematic and pupils are graded on the basis of rather loose 'grade descriptions' while the proportions achieving each grade were held roughly constant in the first two years in line with the previous public examination. Since the announcement of the national assessment proposals the search for better criterion referencing for GCSE has been halted.

Ironically, one of the justifications for making GCSE criterion-referenced was that it would help to raise standards: since there would be no limit on the number of pupils able to gain top grades this would encourage teachers and pupils to aim, and achieve, higher. The percentage of the age group gaining the top three grades has in fact risen, with the result that there are claims now being made that the exam is too easy. Coursework assessment is also seen by the administration as being not sufficiently rigorous and too dependent on teachers; as a result, for most subjects, a maximum of only 20% of the marks may now be awarded for coursework.

Since the introduction of the national curriculum into schools, two changes have been announced to it which have resulted directly from the difficulties of aligning the GCSE with the national curriculum and assessment programme. Since each GCSE course requires 10% of curriculum time (for the two years from 14–16) it is clear that not all students could follow a GCSE course in all ten national curriculum subjects (plus religious education) since this would in theory leave no time for other non-statutory aspects of the curriculum (for example, classics, a second foreign language, personal, social and health education). At the beginning of 1991 the secretary of state for education thus announced that the full national curriculum would only be followed to the age of 14. From 14–16 all pupils must follow a full GCSE course in the core subjects (English, mathematics and science); all pupils must study technology and a modern foreign language but not necessarily to GCSE level; all pupils must follow a course of either history or geography or half of each; only a full course will be examined by GCSE. Art and music will be optional at this stage as will physical education, although schools are expected to encourage all pupils to continue with some form of the latter. Subjects which are not automatically assessed via the GCSE (i.e. all except the core) may be assessed via examinations developed by the vocational examining bodies. The expectation is that higher achieving pupils will take GCSEs while lower-achieving pupils will go for more vocational qualifications. Thus the notion of a full entitlement curriculum for all, offering a broad general education to 16, has been watered down: the 'option' system at 14 is now again similar to that which was already operating in many schools before 1988. Although all pupils must now continue with a full course of science and some technology to 16, an academic/vocational divide continues to be built in to 14+ curriculum provision.

The second major change to come about is a restructuring of the maths and science curricula. The original national curriculum structure gave mathematics 14 attainment targets and science 17. The examining bodies which are responsible for producing, selling, marking and analysing the GCSE announced that they could not report performance on the ten-level scale in relation to this AT structure. As a result, both curricula have been streamlined to five (broader) ATs with approximately half the number of statements of attainment, while the programmes of study remain largely unchanged. This new structure should therefore not affect teaching plans but will make the assessment simpler for both teachers and examining bodies.

It is a clear indication of the perceived importance of the GCSE that its requirements were allowed to modify the national curriculum and assessment programme in this way rather than vice versa. Furthermore, the return towards the domination of the formal written examination mirrors developments in relation to the SATs. This will be discussed further in the following section.

Towards a new assessment paradigm: graded tests and records of achievement (RoA)

Developments in assessment in England and Wales over the last five to eight years have attempted to embrace a range of new approaches: criterion referencing, teacher-based assessment, active process-based assessment tasks, and coursework assessment. This shift in assessment paradigm from a broadly psychometric, norm-referenced, examination-based model towards an educational assessment model is well illustrated by the philosophy outlined by the TGAT report. As already suggested it said that TA should be a fundamental element of the system and that the information generated by assessment should serve several purposes: formative; diagnostic; summative (to record the overall achievement of a pupil in a systematic way); and evaluative (so that aspects of the work of a school could be assessed). The report was acknowledged as being far sighted, professionally supportive and likely to encourage good practice in assessment and teaching. There were, however, criticisms from some educationalists of the ten-level system. There were concerns over the extent of external testing, the playing down of TA in relation to SATs and the publication of unadjusted national assessment results as a basis for school accountability.

The moves towards criterion referencing, continuous assessment based on teacher judgement, and active or extended assessment tasks are common to GCSE and national assessment. The latter two elements are time-consuming for teachers but are seen as contributing to their professional role. Where the teacher-based assessments are linked with external assessment and/or reported as part of a certification procedure, external moderation is involved, which is also time and resource consuming, but can again act to promote professional development.

Other developments in assessment practice in England and Wales have also shared these tensions. Among the most significant are graded assessments and RoA. Graded assessment developed from attempts to modularize the curriculum and to offer pupils shorter-term goals and individual rates of progression through the curriculum.

The assessments themselves could be either classroom based or examination based, although the latter was difficult to reconcile with the notion of readiness—taking the assessment *only* when ready to pass—which is a crucial element in the argument for the motivating properties of graded assessment. Graded assessment has been a popular development at secondary-school level, in maths and modern languages in particular, and teachers have reported increased student motivation, notably among lower achievers (Pennycuick and Murphy, 1988). There are, however, organizational problems relating to its management and flexibility, and technical problems relating to the hierarchical ordering of material and the grade descriptions of the certificates awarded. Problems of developing statements which express unambiguous hierarchies of attainment, the level of specificity of criteria and the generalizability of performance beyond the context of the assessment are the same as those raised by other forms of criterion-referenced assessment. A number of graded assessment schemes were made equivalent to GCSE, but since the edict about the amount of terminal examination assessment required for GCSE their importance has significantly declined. Graded assessment is built on a model of learning which requires learners to have clear information on learning objectives and regular feedback; thus it is more interactive than the traditional secondary-school examination course and emphasizes the use of assessment primarily to support learning though providing for certification as well.

This is arguably even more true of the assessment model underlying profiles and RoAs—another major assessment policy development of recent years. The aim is to record the broad range of a pupil's achievements—academic and personal—on both a formative and a summative basis. As such, RoAs are rather more interactive and dynamic though sharing many features with graded assessments. RoAs also involve the explicit stating of objectives which are discussed and negotiated with the pupil. This dialogue with pupils includes reflections on their attainment, and through dialogue, pupils are encouraged to take more responsibility for their own learning. The content of the RoA is also wider than the narrowly academic: it is an attempt to provide more comprehensive, constructive and meaningful records of pupils' achievement in school, emanating from an era in which public examinations were aimed only at the top 60% of the ability range. 'Profiling' is the procedure in which pupils and teachers jointly construct an assessment record over a wide range of academic and personal objectives. The RoA is the summative document which results from the profiling process and which pupils have when they leave school or college. The assessment processes on which RoAs are based are sometimes known as descriptive reporting or assessment, and the limitation, as far as accountability or evaluative procedures are concerned, is that the descriptions are not amenable to numerical or grade-based summarizing. Indeed the proponents of RoAs would be against such a move since modifying the summative document to produce quantitative descriptors would jeopardize the nature of the profiling process and the centrality of the formative, teacher–pupil interaction.

RoAs have been essentially a grass-roots development among schools and teachers, mostly at secondary level. In 1984 the government issued a policy document which expressed its commitment to providing RoAs for all school leavers by the end of that decade. In 1990 the DES regulations on reporting pupils' achievements called for RoAs, but the requirement was simply for a 'document of record' and did not embrace the profiling process leading up to it. Whilst the formative process is not explicitly discouraged, the absence of any specific obligation on schools to provide for something that is both time-consuming and competing with a demanding new range of legally required activities, means that much of the progress made in terms of developing supportive formative assessment practices during the last decade as part of the RoA initiative has now been lost. Nevertheless, since 1993 all school leavers have been issued with a national Record of Achievement which includes information on their qualifications, evidence about their skills, their attendance rate, their success in non-academic spheres and a personal statement by the student.

Another very significant assessment development in recent years as far as schools are concerned is the advent of GNVQs. Designed to introduce students to a broadly-defined vocational area such as 'health and social care' or 'leisure and tourism', the design of these courses puts a great emphasis on competency-based assessment. Students must show, in the various 'units' that they complete for the course, that they have covered all the performance criteria set within a range of settings. Furthermore, great emphasis is put on the demonstration of key skills such as 'action planning' and evaluation. A range of core skills such as communication and problem solving are also built into the assessment of the units, all of which are assessed by the teacher. In addition, students must pass some multiple-choice-based tests of knowledge. One of the most significant features of GNVQs is their largely criteria-referenced approach. Originally they were concerned with a pass/fail qualification but it was found necessary to introduce a minimal level of differentiation for higher education selection purposes so it is now possible to gain a 'merit' or 'distinction' award by demonstrating more highly developed skills coupled with more substantial and independent work. The assessments made are verified both internally and externally but it is generally acknowledged that the adoption of such radically new continuous assessment procedures for what is now a very large number of candidates is associated with many problems of manageability and reliability (GARP, 1994; SAGE, 1995). Nevertheless, the capacity of GNVQ courses to attract large numbers of students to stay on in full-time education may be traced at least in part to its more learning-centred assessment arrangements.

The GNVQ is the creation largely of the National Council for Vocational Qualifications and has been sponsored in its development by the Department for Employment rather than that for Education. The merger of these two government departments in 1995 thus brings together two very different sets of assessment priorities and traditions, the former emphasizing forms of assessment which encourage the motivation to continue learning and the acquisition of vocationally-relevant skills; the latter, as we have seen, clinging to the security and assumptions of tradition. It remains to be seen whether the GNVQ will prove sufficiently powerful to provide an assessment Trojan horse.

Policy issues

The GCSE with its certificating role is a classic example of a 'high stakes' assessment, and it is clear that it has had an effect on curriculum and pedagogy. The national assessment SATs for England and Wales are also 'high stakes' (since pupils, schools and possibly teachers will be evaluated on the basis of results) and there is evidence that the style and content of the early SATs for seven-year-olds influenced infant teachers' practice (NFER/BGC, 1992). In both these cases many of the moves are towards what educationists would regard in the main as better practice: a move away from restrictive teaching and learning styles and, at seven, towards more work with small groups of children (Gipps *et al.*, 1992). In both cases, also, the central role of the teacher in the assessment process has contributed to their professional development and engagement.

Assessment trends however are in the process of reversal: the government is not in favour of coursework assessment, time-consuming performance-based SATs, or TA dominating at certificating or reporting stages. The move is therefore back towards the domination of traditional examination procedures and paper-and-pencil exercises with all that this will mean for classroom practice. That said, the traditional examination procedures are not of the multiple-choice type but allow for assessment of extended essay writing and higher order thinking skills.

The feasibility and effect of working to a national curriculum which is structured in terms of a defined progression of teaching and learning, with its underlying concept of linear progression which is at odds with constructivist models of learning, has yet to be judged. The effect of having high status external assessment in only the core can be predicted, yet the fact that the rest of the curriculum is legislated may soften the effect. It is, however, a significant reversal of the move towards an educational model of assessment, and it is important to ask why this has happened.

Assessment is being used by this administration, as by many others, to gear up the education system, to raise standards and to force accountability on schools. In this climate teachers are not to be trusted as their own evaluators. Neither are 'elaborate, time-consuming' assessment tasks considered appropriate. The formal, unseen examination is regarded as having served the system well in the past and, it is presumed, will do so again. Politicians see it as more objective, more reliable and cheaper. It is also felt by many traditionalists that the more open relationship between teacher and pupil, which is a strength of the RoA movement, for example, is inappropriate.

Apart from the appeal of the tried and tested, there are two other issues which have contributed to the reversal of fortune for a more educational emphasis in the assessment practices of England and Wales. In the case of national assessment one of the most significant problems may be found in the TGAT model itself. In the TGAT report there was little mention of standards and how these could be raised by testing, and limited emphasis on accountability procedures. The tone of the report was thus at odds with the political climate within which the national curriculum and assessment were introduced. Small wonder then that, as teachers complained of the workload involved in SATs and the low level of standardization became clear, it was quickly decided that the 'SATs' for 1992 would be largely paper-and-pencil tests, standardized, and capable of being taken by the whole class at once. In addition, such performance-based assessment is an approach that is not suited to surveying the performance of every pupil of a particular age group at a certain point in time, particularly given the complex structure of the national curriculum to which it is linked. It is simply too time consuming, and if both the TA and the SATs have to be moderated externally in order to provide for comparability the task becomes even more daunting.

The national assessment blueprint thus did not support the administration's requirements. Add to the problems of implementing the national assessment model, the apparent lowering of standards in GCSE and it was not surprising that the government felt justified in calling a halt to these particular educational developments even though the evidence concerning their very positive effect on learning was steadily mounting.

Performance-based assessment appears not to be appropriate for assessing literally hundreds of summative criteria. It is, on the other hand, ideal for supporting individual, formative and diagnostic assessment by teachers for their own purposes. Although continuous assessment by teachers can well be aggregated for summative purposes, the English tradition of external control through assessment tends to support the view that summative assessment, particularly if it is also to be used for evaluative purposes or for certification and selection, must be taken out of the hands of teachers. It is of course true that teachers do need some form of referencing if their standards are to be comparable across the country, which fairness and equity demand. At GCSE and A level, external markers and moderation processes have been developed to deal with this issue and are widely accepted as producing reliable judgements. However, an assessment system which relied on widespread moderation and extended marking when applied to four age groups simultaneously would clearly be unmanageable. And in any case, as Linn *et al.* put it '... if great weight is attached to the traditional criteria of efficiency, reliability and comparability of assessments from year to year, the more complex and time-consuming performance-based measures will compare unfavourably with traditional standardized tests' (Linn *et al.*, 1991).

A core problem of any assessment system is likely to be the incompatibility of its various purposes. It has been argued that the original TGAT model, for example, suggested that the same system of assessment could serve all the required purposes for a scheme of national assessment—formative, diagnostic, summative and evaluative. The notion that one programme of assessment could fulfil all four functions was always questionable and has been shown to be false: different purposes require different models of assessment and different relationships between teacher and pupil. Assessment for formative purposes is essentially carried out by the teacher in an informal way, often with no clear conclusions. The repeated assessment at an informal level allows the teacher to form valid judgements of the pupil's performance particularly because they are able to assess the pupil in a number of settings and contexts. External assessment for summative and evaluative purposes tends to be one-off and external to the teacher–pupil relationship. Assessment information collected formatively by teachers, when summarized can be unreliable and if used for the purpose of accountability or quality control may well be unfair. Furthermore, its use for this latter purpose in turn severely impairs its formative role.

Conclusion

There are three lessons to be learnt from recent assessment developments in England and Wales. The first is that good quality assessment is time consuming and requires resources and professional training to be committed to it by policy makers if it is to succeed. The second is that the two general functions of assessment—the formative and the summative—are difficult to reconcile. Last and perhaps most importantly it is clear that assessment frameworks are likely to be heavily influenced by particular policy traditions. Those initiatives which are not in harmony with the aims and understanding of the government of the time are unlikely to survive.

In England and Wales, these lessons have been learned the hard way. The enthusiasm and energy that characterized the growth of radically new assessment approaches in the 1980s has been squashed by the powerful reassertion of traditionalist assessment assumptions which emphasize the use of results rather than the support of learning. Until the government in England and Wales finds some other, more powerful way of exerting leverage on the education system, this reliance on assessment to control the system is likely to continue. Equally it may be that the pressure from the world of employment will prove an even more powerful agenda and overthrow this tyranny of tradition.

References

Brown, M. (1992). Elaborate nonsense? The muddled tale of SATs in mathematics at key stage 3. In Gipps, C. (ed.) *Developing Assessment for the National Curriculum*. Bedford Way Series. London: Institute of Education/Kogan Page.

Gipps, C., McCallum, B., McAlister, S. and Brown, M. (1992). National assessment at 7: some emerging themes. In Gipps, C. (ed.) *Developing Assessment for the National Curriculum*. Bedford Way Series. London: Institute of Education/ Kogan Page.

Jennings, A. (1992). Seeing the wood for the trees: the assessment of science at key stage 3. In Gipps, C. (ed.) *Developing Assessment for the National Curriculum*. Bedford Way Series. London: Institute of Education/Kogan Page.

Linn, R. L., Baker, E. L. and Dunbar, S. B. (1991). Complex, performance-based assessment: expectations and validation criteria. *Educational Researcher*, **20**(8), 15–21.

Pennycuick, D. and Murphy, R. (1988). *The Impact of Graded Tests*. London: Falmer.

Prais, S. and Wagner, K. (1983). *Schooling Standards in Britain and Germany*. London: National Institute for Economic and Social Research.

Silver, H. (1979). Accountability in education: towards a history of some English features. Paper for SSRC Seminar on Accountability, June.

Stobart, G. and Burgess, T. (1992). Assessing English at key stage 3: dilemmas for SAT developers. In Gipps, C. (ed.) *Developing Assessment for the National Curriculum*. Bedford Way Series. London: Institute of Education/Kogan Page.

CHAPTER NINE

Shifts towards monitoring in the world's largest education selection system: current Chinese experience

Meng Hong-wei

The formal education system in China

The structure of the Chinese formal school system has changed over time, but the basic structure of the present system was established in the 1950s. The general picture of the formal school system in the 1990s is presented in Fig. 9.1.

China has the largest formal education system and school population in the world. By the end of 1991, the total enrolment was almost 200 million. Schools enrolled 22 million preschool children, 121.6 million primary students, 53.7 million secondary students and 2.04 million higher-education students (State Education Commission, 1992: 5). The system is highly selective. Table 9.1 shows the estimated flow of a cohort of students from admission to Grade 1 primary in 1977 to admission to university in 1989. Only around 2% of those who entered primary could expect to enter university.

The pyramid structure of enrolments indicates that educational selection occurs at several critical points. In areas where access to junior secondary schools is restricted there is selection at the end of the primary cycle. In all areas there is selection at the end of junior secondary, and again at the end of the senior secondary cycle. Limited resources and a huge population in China mean that less than 3% of an age cohort of students succeed eventually in gaining admission to the university. This situation is

Age of entry   Grade

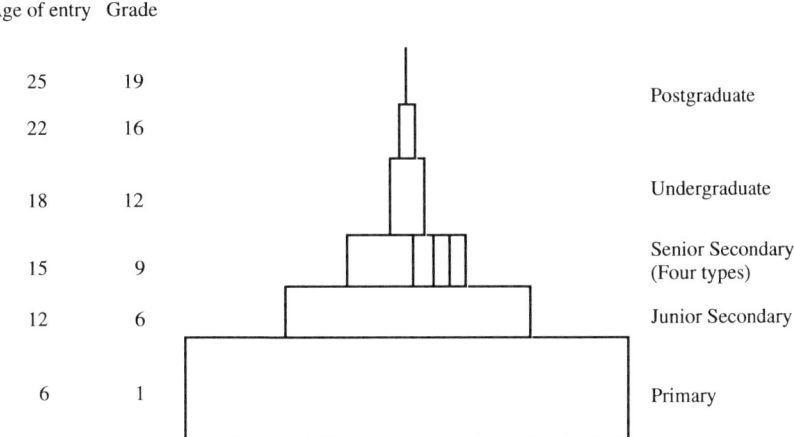

Figure 9.1   The structure of the education system in China.

Table 9.1   Flow of cohort through the education system

Year	Grade	School enrolled	Enrolment	Percent
1977	1	Primary	31,115,000	100
1983	7	Lower Secondary	13,171,000	42
1986	10	Upper Secondary	4,378,000	14
1989	13	University	597,000	2

Source: Lewin and Little (1993).

likely to continue for a long time. The implementation of a nine-year compulsory education cycle has made selection less important in the transition from primary to junior secondary education, but extremely important from junior secondary to senior secondary; and from senior secondary to higher education. At the primary and junior secondary stages of education, the role of student assessment is beginning to shift from selection toward the monitoring of learning achievement. It is this recent emphasis on monitoring which provides the focus of this chapter.

Background to the shift in direction from selection to monitoring

Educational quality and the university entrance rate

After the Cultural Revolution, Deng Xiao Ping, as well as a growing number of other high-ranking leaders, called for a new emphasis on education, science and technology. The quality of education declined during the Cultural Revolution and led to the lack of a qualified workforce, affecting industry and hampering the process of modernization.

In a May 1977 talk to members of the Central Committee, Deng had set the tone for the future policies on modernization and education:

> it is necessary to improve education at every level ... to promote scientific and technological work ... standards of education [must be raised] at the same time as we make it available to more and more people. (Deng Xiao Ping, 1984: 49)

The Ministry of Education, acting in the spirit of the instructions of the Communist Party Central Committee, adopted a series of concrete measures to restore the educational methods proven effective before the Cultural Revolution (Zhou Yu Liang *et al.*, 1990). These included:

- restoration of 'the National Unified Entrance Examination for Institutions of Higher Education' to ensure the standard of the students in the universities
- reapplication of the 'key school' policy.
- redevelopment of the syllabus and rewriting of teaching materials.

At this stage, the major concern in the raising of quality was the 'bringing of order out of chaos' (*bo luan fan zheng*) in all educational enterprises, but the proposals to improve 'quality' were rather diffuse with little guidance on the definition and measurement of 'quality'.

The proportion of graduates who pass the tertiary entrance examination (the National Unified Entrance Examination for Institutions of Higher Education) and enter a university or college has always been considered as a measure of the quality of teaching in secondary schools by parents, teachers, and students themselves. Although the examination was abolished during the cultural revolution, it was restored in 1977. A few years later, it became apparent that the desire for success in the tertiary entrance examination had come to dominate the teaching and learning activities in secondary schools. As Fensham (1987: 71) has noted, 'a textbook may include a lot of content ..., but if it is not examined then it will be perceived by learners as having little worth'. As a result, the structure of knowledge learned at secondary school is seriously biased. It is strongly influenced by the examination. This is similar to the situation found in several Asian countries (Oxenham, 1984; Dore, 1976). In China, the examination system is called 'the baton' (*zhi hui bang*), directing the teaching and learning at which it points. The baton can even affect those students studying at the junior secondary and primary levels.

One result of regarding the tertiary entrance examination as a measure of quality of secondary teaching is that schools try to achieve as high a pass rate as possible. This is reflected in the Chinese saying, 'thousands and thousands of soldiers marching across a single-plank bridge' (*qian jun wa mar zhen guo du mu qiu*). The problem is that this march towards examination success violates the principle of 'all-round development', which also underpins contemporary Chinese aims in education.

In 1989 the former president of the State Education Commission, He Dong Chan, stated that the tertiary entrance examination had become the marker of graduation from senior secondary schooling, the consequence of which was that many secondary students appear to leave school as failures. In fact most of them are qualified graduates, representing precious wealth of the nation. Because there is no official recognition of their achievements at the end of senior secondary school, it seems that among the two and half million senior secondary school graduates only 600,000 are victors (He Dong Chan, 1989).

The former vice-president of the State Education Commission, Zou Shi Yan commented that

> while qualified freshmen for universities are needed, the needs of the majority of students who cannot enter university cannot be disregarded. Nor can university entrance be used as a criterion for evaluating student and school performance. (Zou Shi Yan, 1991)

In order to establish a scientific and systematic educational quality monitoring system at all levels to take the place of the tertiary examination, the report on the seventh five-year plan at the fourth plenary of the sixth National People's Congress specified that the administration of educational institutions should be enhanced, gradually establishing a systematic educational evaluation and supervisory system (Zhao Zi Yian, 1985: 22).

As the government began to consider the establishment of a monitoring system, Liu Bin, vice-president of the State Education Commission, stressed the need

> to establish a set of scientific and reasonable standards evaluating the quality of the basic education ... [and] ... taking the reform of examination system as a breach, to perfect the evaluation of the quality of primary and secondary schooling is one of the important reforms in basic education. (Liu Bin, 1989)

One of the proposed measures was the introduction of a certificate examination for students in senior secondary schools. The examination uses a unified standard (criterion) to assess whether a student has mastered all the basic knowledge in senior secondary school and the standard achieved. The main task of this system is to ensure the quality of teaching in senior secondary schools, and to provide more information about students' learning and the level achieved. The introduction of this senior-secondary certificate examination system is an attempt to separate the selection examination from the quality-monitoring examination.

Abolish the 'key schools' at primary and junior secondary levels

Key schools are specific to China, and are justified by the idea that it is impossible to run all schools well. Paying more attention to some schools with better conditions and giving extra supplements was seen as a way of solving the problem of quality during the Cultural Revolution. After the Cultural Revolution, an urgent requirement was to alleviate the shortages of a qualified workforce. The aim of the present key secondary schools is the use of limited physical and human resources to raise the quality of the teaching in schools and to foster more qualified specialists (Ministry of Education, 1984: 168).

Criticism and debates on the policy of key schools have continued for many years. A major issue is the creation of a gap between two types of schools, generated by the key-school policy. In the ordinary schools students have less positive attitudes toward school life. Teachers and parents become demoralized because their students have less opportunity to enter tertiary education than students enrolled in key schools.

In the late 1980s the State Education Commission stipulated that in those parts of the country where access to junior secondary schooling is universal, the former entrance examination to junior secondary schools should be abolished and key junior secondary schools discontinued. Students were to be assigned to junior secondary schools based on the region in which they live (State Education Commission, 1988). Abolition of the entrance examination to junior secondary schools was seen to demand a system of monitoring and evaluating the quality of primary schools and their performance.

Educational efficiency

In 1979 a growing awareness of the need for higher efficiency in economic fields appeared in the media in China (Yeh, 1984). This was the starting point for emphasizing the efficiency of the production process after the Cultural Revolution. Chinese economists modelled an input–output relationship between the quantity and quality of products and the cost and consumption of materials in the production process. Their aim was to achieve greater efficiency through less inputs and more profits. Education is the largest sector in the national economy in China. The discussion of greater economic efficiency was quickly transferred to education in the belief that the analysis of the efficiency of the investment in education was an important guide to improvement in education. Although most of these studies were developed by economists, the improvement of educational quality and efficiency has became a major preoccupation among educational administrators and politicians. The more education is recognized as an investment in human resources and as an investment for bringing about economic growth and social change, the stronger becomes the need to increase the efficiency of the educational system. For an effective educational system, the adequate monitoring of the quality of primary and secondary schools requires detailed, specific and reliable information on student learning achievement. It is desirable that the educational accomplishments of students are as high as possible with respect to the stated aims.

In 1980 and again in 1984, a team of experts from the World Bank visited China, and reported that China's achievement in education since 1949 had been unmatched among developing countries of the same income level. Enrolments in formal and non-formal primary and secondary education were high by any standards. The supply of teachers and teaching materials were good in many respects and student performance quite high as far as could be assessed (World Bank, 1985: ix). The experts also pointed out, however, that if the move to universal nine-year basic education was to make the required contribution to China's economic and social development, much time-consuming work was needed in several areas. They indicated that a complete analysis of China's basic education system would require:

- a survey of physical facilities
- a large-scale data collection survey using carefully designed sampling techniques leading towards an input–output analysis (IEA type study)
- a provincial or even prefectural breakdown of school enrolments by grade, sex, ethnic group, and rural/urban residence as related to age group and the composition and distribution of the population. (World Bank, 1985: 17)

These comments were profound and indicated the need for a system to monitor quality very different from that of the past, where the only criterion used to assess schools was the proportion of graduates entering tertiary institutions.

The improvement of teaching conditions in compulsory education

The 1985 Compulsory Education Act has led to an improvement in basic education through investment in school building and basic infrastructure. In 1991 the Chinese National People's Congress appointed its Education, Science and Health Committee to lead an evaluation of the implementation of the Compulsory Education Act. Fifty percent of the provinces, autonomous regions and municipalities achieved the goal of 'no dangerous room in every school, a classroom for each class and a study desk and chair for each student' (*yi wu lian you*) over the five-year period since the introduction of the Act.

Liu Bin, vice-president of the State Education Commission, has stated that the average expenditure on each primary and secondary student has been raised gradually between 1985 and 1990 because of the efforts of local government (Liu Bin, 1991). In 1989 the average cost per primary and secondary student was 91 and 212 RMB Yuan, 194% and 165% of the 1985 figures, respectively. In the regions in which the junior secondary school has been universalized, 70% of cities have abolished the selective entrance examinations to junior secondary schools. By 1990, the percentage of graduates of primary school admitted into junior secondary school was 74% (Liu Bin, 1991). Net enrolment rates in 1990 had reached 97.83% across the country, the figure in rural areas being 97.4%. The retention rate of enrolment in primary schools had reached 98.4%. With a national net enrolment rate of 98%, and improvements in teaching conditions, one of the major challenges facing universalized compulsory education is the attainment by all pupils of specified learning outcomes. In February 1993 the Central Committee of the Chinese Communist Party and State Council issued a National Programme for Educational Reform and Development in China. The programme aims to expand education and increase the quality of education and effectiveness of schooling. In the process of development of basic education, school quality has to be improved gradually and standardized. Schooling at primary and secondary level should shift its orientation from examinations to all-round development aimed at enhancing the quality of the nation.

The shift in assessment from selection towards monitoring

To establish a supervision system

The supervision system in China was set up by the State Education Commission in 1986 in order to support the implementation of the Communist Party's educational policy. In order to supervise the development of educational quality across the country, guidelines to quality were specified in more detail, as outlined in Fig. 9.2.

These broad guidelines have been used to create a more detailed specification of each item in operational terms by the local supervision authorities. The guidelines are used particularly for primary school students when they finish their study in primary school. This assessment is used for monitoring rather than selection in areas where access to junior school has been universalized. An example of a detailed specification of the guidelines on the moral dimension of quality, taken from a Beijing primary school, is presented in Fig. 9.3. The specifications differ slightly for different grades.

Methods of assessment have also been specified in the handbooks, e.g. 'according to the requirement set by the moral syllabus, the teacher should use the 100-score system to mark each student through interview, oral exam, questionnaire or pencil paper test at the end of a term'.

---

1. **Moral**
   Attitude towards party's and government's policy
   Ideology and moral character
   Behaviour and habit

2. **Intelligence**
   Basic knowledge
   Basic skills

3. **Physical culture, health**
   Growth, physical stature of students
   Pass ratio of physical culture
   Physical and health knowledge
   Physical sports and health habits, smoking status of students in the school
   The proportion of students with eyesight problems and other common diseases

4. **Aesthetic education**
   Basic knowledge of music and painting
   Appreciation of the beautiful and aesthetic judgement

5. **Labour education, work skills education**
   Attitude to labour, labour habits
   Basic knowledge and skills on labour
   Ability to take care of oneself in living

6. **Students' interests and special skills**

7. **The evaluation of the graduates from society, employers and higher level schools or universities**

Figure 9.2   Guidelines to educational quality.

---

1. Moral
1.1. Love the motherland, love people; take good care of national flag, emblem; be able to sing the national anthem; love father, mother, brother, sister, teacher, classmate; love school, home town, the capital of the country and all mountain and rivers in the country, be willing to learn for the construction of socialist motherland.
1.2. The basic knowledge of ideology and moral character; achieve the requirements of the moral syllabus.
1.3. The assessment scale for ideology and moral character. The basic requirements are:
   • know himself/herself to be a Chinese. Know the whole name of the country, the capital and the state leaders.
   • know the national flag, emblem, map etc; be able to sing the national anthem.
   • understand that he or she should give a salute when the national flag is being raised and stand as the band strikes up the national anthem.
   • know how many important holidays there are in China.
   • study hard and become a good student.

Figure 9.3   Detailed specification of the 'moral' dimension of educational quality.

But there is currently a weakness in the assessment of student learning achievement. The only assessment mentioned is the formative test and term-end examination, both of which are school-based. The pedagogical divisions of the local educational bureau are concentrating their efforts in this area. Pedagogical divisions in the local educational bureau have been in place since the 1950s. In 1955 the Ministry of Education issued a document on enhancing the quality of primary and secondary schools (Ministry of Education, 1955). Although these divisions had lapsed during the cultural revolution, each province, prefectural and county had re-established them by the 1980s. They advise local education authorities on teaching methods and quality control (Beijing Educational Administration College, 1983). One of the tasks of this division is to gather the data on the student learning achievement in each subject.

The pedagogical research system

To assess the student learning achievement, the pedagogical research system across the country has played a very important role since the 1950s. Pedagogical research divisions at all levels control teaching quality. For each subject there is a supervisor from the education department who reviews the teaching plan and summary writing by a school teacher; and reviews the test paper, visits schools and engages in discussions with teacher or student.

The most frequently used means of assessing student achievement has been sample surveys of achievement in specific subjects. The pedagogical research system in the contemporary education system in China is shown in Fig. 9.4.

In the 1980s the theory of mastery learning and the teaching taxonomy created by B. S. Bloom were introduced to China through the educational journals and newspapers. Chinese educators and educational researchers, particularly those in the pedagogical division at all levels, showed great interest in them. Many tried to apply the taxonomy to student assessment and to monitor educational progress at the local level. In classroom teaching, this was done by setting a learning target using the taxonomy and testing students to see whether they have achieved the set learning target. This has become very popular in primary and secondary schools, and is

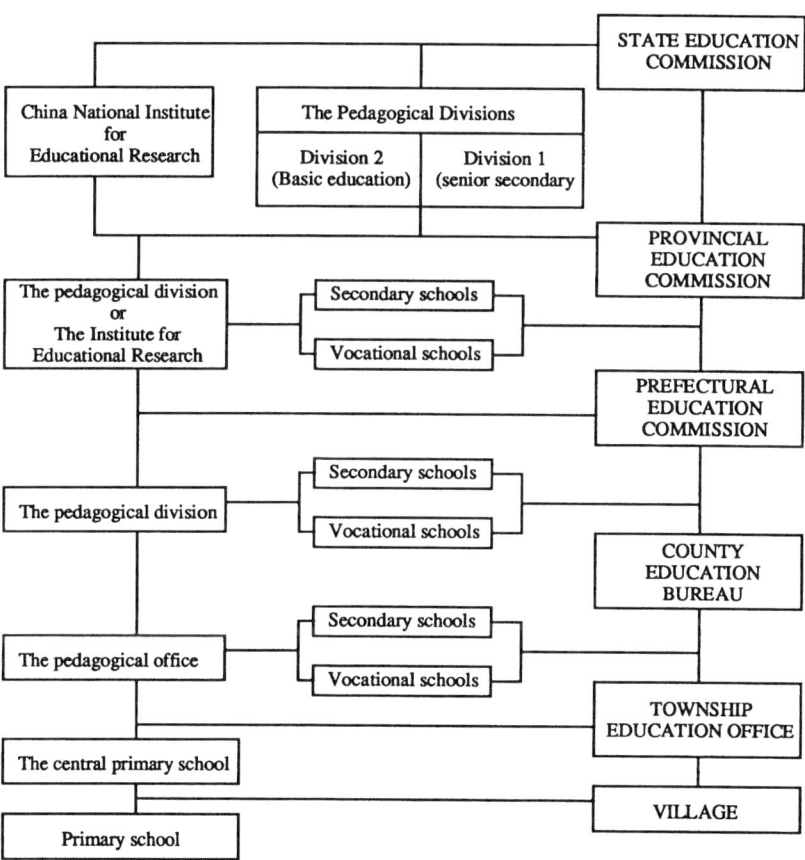

Figure 9.4 The administration and pedagogical system in the contemporary education system in China.

usually guided by the local pedagogical divisions. For example, in Jiangsu province some of the county pedagogical divisions have conducted research to specify the targets for each lesson based on the curriculum guide stipulated by the State Education Commission.

According to these teaching targets, the pedagogical division also provides a set of test items for diagnosis of the students' learning achievement. Teachers use the assessment to improve their teaching through using these tests and forms.

The certificate for examination of general senior secondary school

From 1985 on, the State Education Commission carried out an experiment in Shanghai. The local education department in Shanghai set a certificate examination for students in senior secondary schools. Five years later, in 1990, the State Education Commission issued a document on the certificate for examination of general senior secondary schools across the country which clearly specified that the aims of the examination are to implement the party's educational policy, to strength quality control, to promote teaching reforms and to raise teaching quality widely (State Education Commission, 1990).

This examination is now a nationally recognized test in general senior secondary schools determined by provincial criteria. It is a means for monitoring and evaluating the quality of teaching in senior secondary schools. The examination uses a unified standard (criterion) to assess whether a student has mastered all the basic knowledge in senior secondary school and to check what standard has been achieved. The main task of this system is to ensure the quality of teaching in senior secondary schools and provide more information about students' learning.

It is the first move in China to separate the tertiary education entrance examination from a certification test for senior secondary schools. The certificates are issued by the province, municipalities and autonomous regions. The graduating students sit a certification examination in nine subjects. Students who pass in nine subjects receive a certificate of examination of the general senior secondary school. The holder of the certificate can then apply to take part in the tertiary education entrance examination. Subjects included in the entrance examination are limited to a group of three subjects which depend on the major subject of study in the university. Grades A–E are awarded in the examination, and grades of D and above are a pass. The local education department intends to strengthen its guidance to schools where weakness in teaching is found in a given subject, according to grades gained by the students. In the selection of university freshmen, achievement in the certificate examination and selected courses, students' behaviour and capability in the interest group will be considered. This means that selection will no longer depend on the scores in a single examination.

The new general senior secondary school examination is welcomed by teachers and students in very poor schools. It does not conflict with the tertiary entrance examination, because traditionally these schools have never experienced success in this selection examination. Now most of the students pass the certification examination and demonstrate their achievement. The key schools, on the other hand, view this examination as extra work, because their target remains the tertiary entrance examination.

The major monitoring programmes in China during 1985–93

The joint study on moral education of youth and children. This project was launched in 1990. The aim of the study is to identify and explore suitable methods and contents of moral education for children and adolescents in China and other countries, especially developing countries. In order to realize this aim, efforts were made to investigate the present state of Chinese adolescents' moral values and moral behaviour, to study the successful experience and the main challenges we are facing in the area of ethical and moral education, and to develop measures for improving moral education.

Data were collected from 'the questionnaires for surveying the situation of moral behavioural codes of children and adolescents' which were based on investigations and pretests in China. An objective sample was drawn from 11 provinces, autonomous regions and municipalities. The sample included 1472 primary 5th Graders, 2279 junior 2nd Graders and 1967 senior 2nd Graders.

The results of the study suggest that:

• moral education is an important way of cultivating students to become qualified citizens
• the education of moral behavioural codes should be included as a formal course in the curriculum of primary and secondary schools
• the content of moral behavioural code education should be adapted to the features of modern times, and western and oriental traditions. (Zhou Yu Liang and Zhang Zhi Yi, 1992).

National survey of students' constitution and health. During 1983–85, under the leadership of the China State Education Commission, the State Physical Culture Commission, the Ministry of Public Health and the State Nationalities Affairs Commission, a national survey of the constitution and health of the Chinese students was conducted in 29 provinces, national autonomous areas and municipalities. A total of 984,872 students, between 7 and 22 years of age, from 2188 universities and schools were involved.

The main purpose of the survey was to describe the present state of student's growth and development across the country and to monitor changes and progress. The investigation included eight health indices, physical morphology characteristics and student performance data in specified physical exercises.

Second IEA science study (SISS). The SISS is an international comparative educational research project conducted by the IEA (see Chapter 4). It attempted to document the state of science education programmes both within each country and across the countries taking part, (Keeves, 1984).

The China National Institute for Educational Research (CNIER) joined IEA in 1984 and took part in the SISS. The purposes of conducting this study in China were twofold: (i) to learn large-scale survey methods used in the IEA survey; and (ii) to look at science education at the secondary level in China in the context of cross-national results (CIER, 1984).

SISS was conducted in China in 1985 with a target population consisting of students in Grade 3 in junior secondary schools in Beijing, Tianjing, and Taiyuan. This corresponds to the definition of population two in SISS. A stratified two-stage probability sample of schools and students was drawn. Data on student performance of science achievement, student characteristics, home background, and teacher and school characteristics were collected from five sets of tests and questionnaires, i.e. cognitive tests, student, teacher, school questionnaires and opportunity to learn questionnaires. SISS was the first attempt to use a large-scale survey as a means of evaluating the performance of the school system in China.

National survey of junior secondary achievement. In 1985 the division of basic education of the State Education Commission, launched a national survey of student learning achievement at the end of junior secondary level. Fifty thousand students from 605 schools in 15 provinces were involved in the survey. The provinces and counties were selected by experts to represent the full range of educational quality. Data were collected on student learning achievement in mathematics, Chinese language and English, together with teaching practices and materials and teacher characteristics in the three subjects (Li Liang You *et al.*, 1988).

UNESCO/UNICEF joint project. The action plan formulated to promote EFA (Education for All) in China envisages strengthening of supervision, inspection and evaluation of the quality of primary and junior secondary education. Although the national supervision system has specified the meaning of educational quality in China, it has not been possible systematically to assess the performance of the primary education system, in the absence of an appropriate mechanism or system for determining the extent to which pupils at different stages of primary education attain the specified learning outcomes. In order to develop an appropriate monitoring mechanism for primary education and to build up a national capacity, the division of basic education of the State Education Commission became involved in the UNESCO/UNICEF joint project on *Monitoring Education-for-All Goals*. The project is in line with the strategies set out at the World Conference on Education for All (WCEFA) which stress the need to focus on learning achievements and outcomes, not simply enrolment.

The target population for this project was specified as the 4th and 6th Grade students in the six-year primary schools. Eight provinces were selected. They are Jilin, Hubei, Shaanxi, Jiangsu, Hunan, Guangdong, Sichuan and Yunan. Twenty-four thousand students from 1200 schools of each grade were involved. The tests of Chinese language, mathematics and selected life skills and a student questionnaire were taken by each student.

The general objective of the study is the development and application of methods and indicators for monitoring learning achievement of pupils at the primary level, on a regular and continuing basis, for the purpose of making available to decision makers information which can form a basis for policies and programmes for qualitative improvement of primary education. The specific objectives of the project are:

- to assess the levels of learning achievement in Chinese language, numeracy and selected life skills of pupils in Grades 4 and 6.
- to determine the extent to which learning achievement of pupils varies across regions/provinces within the country
- to determine the influence of certain pupils, teachers, and school factors, especially the school input factors which could improve the performance of pupils and schools in different area/regions.
- to strengthen national and provincial capacity for periodical monitoring of learning achievement.

These programmes are summarized in Table 9.2.

Problems in the coming decades

The establishment of national standards

The necessity of establishing a nationwide monitoring system is apparent. One problem which arises is the specification of a national standard for

Table 9.2 The monitoring programmes in China

Subject	Time	Target population	Objectives	Sample	Sponsor
Joint study on moral education of youth and children	1991–93	Grade 5 in primary, Grade 2 in lower secondary, Grade 2 in senior secondary	The status of moral values and moral behavioural development	Non-probability typical sample (5718 students)	Badi Foundation
National survey of students' constitution and health	1985	7–22 years	The physical shape, function, quality of youth and children	Multistage random sample (984,872 students)	State Education Commission; State Survey Commission; Physical and Sports Ministry of Public Health; State National Affairs Commission
The Second IEA Science Study	1985	14 years, Grade 3 junior secondary	Learning achievement in physics, chemistry, biology and geography	Two-stage random probability sample (2817 students)	The State Education Commission
National survey of junior secondary achievement	1985	Grade 3 junior secondary	Learning achievement in mathematics, Chinese language, foreign language (English)	Representative sample of 50,000 students	The State Education Commission
UNESCO/ UNICEF joint project	1993	Grades 4 and 6 primary	Learning achievement in mathematics, Chinese language and life skills	Multistage probability random sample (48,000 students, 1200 schools, eight provinces)	UNESCO/UNICEF, the State Education Commission

student learning achievement. In 1992 a comparison among the provinces in terms of the graduate examination pass rate of junior secondary schools showed clearly the disparities among the provinces. How do we establish national standards which can be satisfied by both the developed and less developed areas? How can we take care of the local needs in student learning as well as maintaining national standards?

Sampling design

In order to ascertain the extent to which educational aims have been achieved in the country, we need a representative sample of the target population in the country. But the execution of random multistage sampling is difficult to implement in a country with limited resources for educational research, a huge population, and difficulties with transportation to remote areas. So the search for a statistically acceptable and realistic sampling design for a nationwide representative sample has been seen as an important aspect of establishing a national assessment system for student achievement (Meng Hong-wei, 1991).

Conflict between monitoring and selection

The examination for certification of general senior secondary schooling has had an important impact on schools and has been welcomed by both teachers and students of the poorer schools in particular. Whereas success in the tertiary entrance examination always eluded them, most students in these schools can now demonstrate their achievement through the certification examination. For certain 'key' schools, however, the tertiary entrance examination remains the target and the certification examination is of lesser, supplementary concern. The introduction of the certification examination has created a second structure of assessment with its curriculum demands non-identical with those of the tertiary entrance examination. Potential conflict between the monitoring and certificating assessment system and the tertiary entrance examination will need to be resolved.

References

Beijing Educational Administration College, Division of School Administration (1983). *The Contemporary Regional Educational Administration in China*.

CIER (1984). Summary of the First Joint Meeting of Advisory Committee and Technical Committee of National Center for IEA, March 1984.

Deng Xiao Ping (1984). *Deng Xiao Ping's Selected Works*. Beijing: Foreign Language Press.

Dore, R. P. (1976). *The Diploma Disease*. London: Unwin Education.

Fensham, P. J. (1987). Changing to a science, society and technology approach. In Lewis, J. L. and Kelly, P. J. (eds) *Science and Technology Education and Future Human Needs*. Oxford: Pergamon Press.

He Dong Chan (1989). Speech on the meeting of adjustment of present syllabus of junior and senior secondary schools. *Adjustment of Present Syllabus of General Senior Schools and Implement of Certification Examination of Senior Secondary Schools*. Edited by The Division of Basic Education, State Education Commission: Chang Chun Press, 1991.

Keeves, J. P. (1984). *The Second IEA Science Study: A World Perspective*. The Hague: IEA.

Lewin, K. M. and Little, A. W. (1993). *Assessment and Selection from Kindergarten to University in the People's Republic of China*. British Council Research Monographs No. 3, Beijing.

Lewin, K. M., Little, A. W., Xu Hui and Zheng Ji Wei (1994). *Educational Innovation in China: Tracing the Impact of the 1985 Reforms*. London: Longman.

Li Liang You, Zhang Ri Sheng and Liu Li (eds) (1988). *A History of English Language Teaching in China (Zhong Guo Ying Yu Jiao Xue Shi)*. Shanghai Foreign Language Teaching Press.

Liu Bin (1989). The certification of examination in senior secondary schools and reforms in basic education. In *Adjustment of Present Syllabus of General Senior Schools and Implement of Certification Examination of Senior Secondary Schools*. Edited by The Division of Basic Education, State Education Commission: Chang Chun Press, 1991.

Liu Bin (1991). The implementation of compulsory education in five years. In *The Practice and Experiences of Universalization of Compulsory Education in China*. Edited by The Educational Research Office, Education, Science and Health commission of National People's Congress: The China Democracy and Legal System Press, 1993.

Meng Hong Wei (1991). *The Structural Model of Factors Related to Student Science Achievement in China*. Doctoral dissertation, University of Hong Kong.

Ministry of Education (1955). A circular on the report-back meeting on the secondary education. Internal document.

Ministry of Education (1984). *The Yearbook of China's Education, 1949-1981*. Beijing: People's Education Press.

Oxenham, J. (1984). *Education Versus Qualifications?* London: Unwin Education.

State Education Commission (1988). The document on discontinued key junior secondary schools. *China Education Daily*, (Zhong Gou Jiao Yu Bao) 9 April 1988, Beijing.

State Education Commission (1990). State Education Commission Suggestions on the implementation of the certificate for examination system of General Senior Secondary schools. (Document).

State Education Commission (1991). The guideline for the supervision and evaluation at primary and secondary schools. In Xiping, T. *The Supervision and Evaluation of Basic Education*. Beijing: Education Science Press.

State Education Commission, Department of Planning and Construction (1992). *Educational Statistics Yearbook of China 1991–1992*. Beijing: People's Education Press.

World Bank (1985). *China: Long-Term Development Issues and Options. Annex 1: China. Issues and Prospects in Education*. Washington, DC: The World Bank.

Yeh, K. C. (1984). Macroeconomic changes in the Chinese economy during the readjustment. *The China Quarterly*, 100 (December).

Zhao Zi Yian (1986). *Report on the seventh five years planning*. Edited by the Office of Standing Committee of National People's Congress Conference. Collection of documents from the fourth conference of the Sixth Chinese National People's Congress. Beijing: People's Press.

Zhou Yu Liang *et al.* (eds) (1990). *Education in Contemporary China*. Hunan Education Publishing House.

Zhou Yu Liang and Zhang Zhi Yi (1992). *A Brief Report on the Joint Study on Education of Moral Behavioural Codes to Youth*. Beijing: China National Institute for Educational Research, (unpublished paper).

Zou Shi Yan (1991). Speech on the national meeting of teaching work of general senior secondary schools. In *Adjustment of Present Syllabus of General Senior Secondary Schools and Certification Examination of Senior Secondary Schools*. Edited by the Division of Basic Education, State Education Commission. Chang Chun Press.

Part three
COUNTRY CASE STUDIES: THE BACKWASH OF SELECTION— CONTRASTING CONTEXTS, COMMON DILEMMAS

Throughout this volume, there is an emphasis on the new demands being made of educational assessment, and the degree to which these can be combined and reconciled. Because selection is the most familiar and oldest function of assessment, there is also, inevitably, a tendency to give it rather less overt attention than newer concerns such as monitoring and the promotion of learning. Nonetheless, in every country, selection remains an essential function of assessment—one without which no one can conceive of the education system operating, and which exerts a constant and pervasive pressure on the nature of the assessment process.

What differs from country to country is the degree of selective pressure experienced by students, and the degree to which countries try more or less successfully to combine selection with other functions. On the whole, as Angela Little emphasizes in Chapter 1, selective pressure is associated with economic factors. Developing countries, where there is intense competition for white-collar and middle-class jobs, experience very severe selective pressures, whereas those with successful and developed economies (such as the United States or Western European countries) have systems which, while they may be quite highly competitive, generally do not place so much pressure, so early, on their students.

However, the nature and level of selective pressures is not simply correlated with economic growth rates or income levels. In some countries, entry into the élite operates largely via certain educational institutions. This is especially marked in, for example, France, with an exceptionally tight and small élite largely educated in a few highly selective institutions. Selective pressures are consequently extremely evident—especially for the ambitious middle classes. In Britain, the importance of educational assessment for élite entry is also enormous, though pressures are somewhat less than in France: in part, perhaps, for the simple reason that the top universities (Oxford and Cambridge) are considerably larger, and take more students, than do the French grandes écoles. In Japan, one finds combined a culture which emphasizes that success depends on hard work, a recent history which has rewarded this, and an educational and élite structure in which a great deal depends on whether one attended the right university (Tokyo). Together these produce a hugely competitive educational structure in which children are aware of examination pressures from a very early age.

The United States, Canada and the Scandinavian countries present a marked contrast to their fellow OECD members in this respect, with assessment for selection present, and indeed important, but creating considerably less stress on young people. In other European countries, such as the Netherlands or Germany, the selective functions of assessment are very evident to all students, since it is vital to win admission to the right sort of (academic or technical) school or training in one's mid-teens; but pressure is less apparent for the highest achievers, because élite entry is less tied to particular educational paths.

The other important factor affecting assessment for selection is the degree of political consensus. If there are serious political fractures in a society the selective functions of education also become the focus of the tensions associated with political tensions and contests. This has been very evident in Sri Lanka, for example, as discussed by Little in Chapter 1, and by Kariyewasam in Chapter 13. However, Sri Lanka is by no means unique in this respect. The very importance of assessment for selection in all societies ensures that any major political tensions between groups will be reflected there. The history of black–white relationships in the US over the last 50 years can be tracked by studying students' access to assessment, and the interpretation and use of assessment results, including, for example, the issue of affirmative action for higher education entry.

The case studies in this section are of countries selected because they illustrate both the continuing importance of educational assessment for selection, and the way it reflects the broader economic and political structure of a country. In South Africa, apartheid produced, as Peliwe Lolwana observes in Chapter 10, the 'most fragmented system in the world'; one which now faces the challenge of change into a unified system for a new, post-apartheid society. Egypt, Sri Lanka and Indonesia all illustrate the tensions associated with selection in developing countries where it can be very difficult to move beyond the 'pure' selective function to the goals of evaluation and monitoring emphasized at Jomtein. Yet they also illustrate the diversity of societies at any given level of development and income, and their varying degrees of success in implementing reform have lessons for other systems. Finally Korea, discussed by Chan-Jong Kim in Chapter 12, provides an example of an 'East Asian tiger', where economic success and educational effort are seen by government and citizens as closely associated, but where there have been consistent and continuing attempts to develop monitoring, broaden the nature of assessment questions (and so affect learning), and widen access. The successes and failures of Korean educational assessment policy are highly instructive for anyone interested in assessment for selection within a successful and rapidly industrialized economy.

CHAPTER TEN

The South African assessment and qualification system: opportunities, problems and challenges

Peliwe Lolwana

Introduction

In South Africa, racially-based assessment arrangements have enforced a series of educational and employment inequalities. Over the years the education system has been a battleground over which struggles about the legislated inequalities have been fought. Comparisons between the privileged system for whites, and that for the most disadvantaged groups reveal gross disparities. For example, while almost all white students (*c.* 98%) are guaranteed 12 years of education, less than 40% of African students can boast of such an entitlement (RIEP, 1992). In addition, the pass rates for African students lucky enough to finish high school are shamefully low. For the majority of African students, the education and training system has failed to provide them with qualifications for either further education or employment.

A new, democratically elected, government representing the aspirations of the nation has been in place since 1994. This new government has a mandate to change the education and training system so that it meets the needs of the majority not just those of a minority. As part of this effort, the government will need to put in place assessment policies that, in contrast with those of the past, will contribute toward the improvement of education and training for all. This chapter therefore attempts to identify some of the key problem areas, challenges and opportunities associated with assessment and certification in the future education and training system.

The current system

The current education system presents some problems for the majority of students, and makes it difficult to have a critical mass of students acquiring qualifications for access to higher education and employment opportunities. Even though it is generally agreed that the current system has to be overhauled, it is important to place the problem in perspective.

Divisions along racial lines

The historical fragmentation in South African education has been documented in several places and many times. It is hard to depict an accurate picture of this fragmentation and confusion, but Fig. 10.1 provides an approximate illustration. It represents the organization of the system at the time of the new government's arrival in power.

On the surface, the educational system looks very decentralized. In reality the system has been highly centralized at the top with the Department of National Education (DNE) controlling the norms and standards regarding syllabi, examinations and certification qualifications (NATED, 1992). Public examinations were set and administered by a number of bodies, but centrally moderated by the South African Certification Council (SAFCERT). Examinations, except for those set by the Independent Examinations Board (IEB) fell under the different education departments that had been racially formulated. This racial arrangement over the years enforced inequalities for the consumers of South African education. For example the Department of Education and Training (DET) qualifications (obtained by Africans in white areas) were generally regarded as inferior to any other qualifications, irrespective of the potential and achievement of the individual. The reverse was true of the Department of Education and Culture certification.

Low achievement and progression rates

Low achievement of students and low progression rates[1] are serious problems throughout the education system—especially for African students. Below Grade 12 (the end of high school), repetition rates account for an unacceptably large proportion of the school population. For example, in 1991, 17% of African sub-A (Grade 1) pupils were repeating the year (Fig. 10.2).

Figures 10.3–10.4 also illustrate the problems of the system: Fig. 10.3 by illustrating drop-out rates for African students, and 10.4 by showing matriculation pass rates for different groups. The highest failure rate for African students occurs at Grade 12, in an examination set by the External

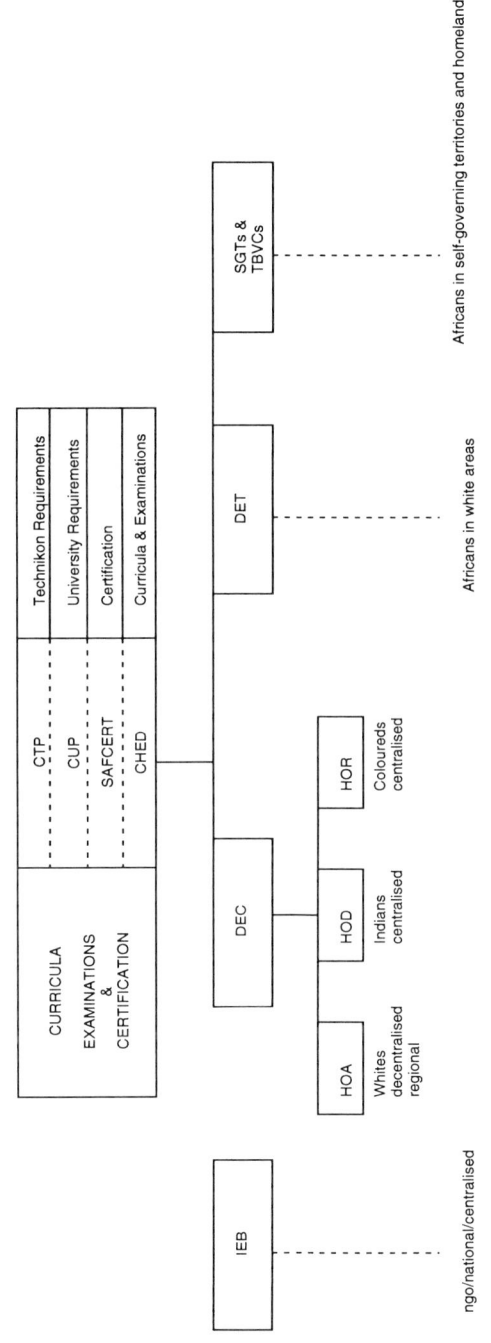

Figure 10.1   The organization of South African education and examinations in 1994.

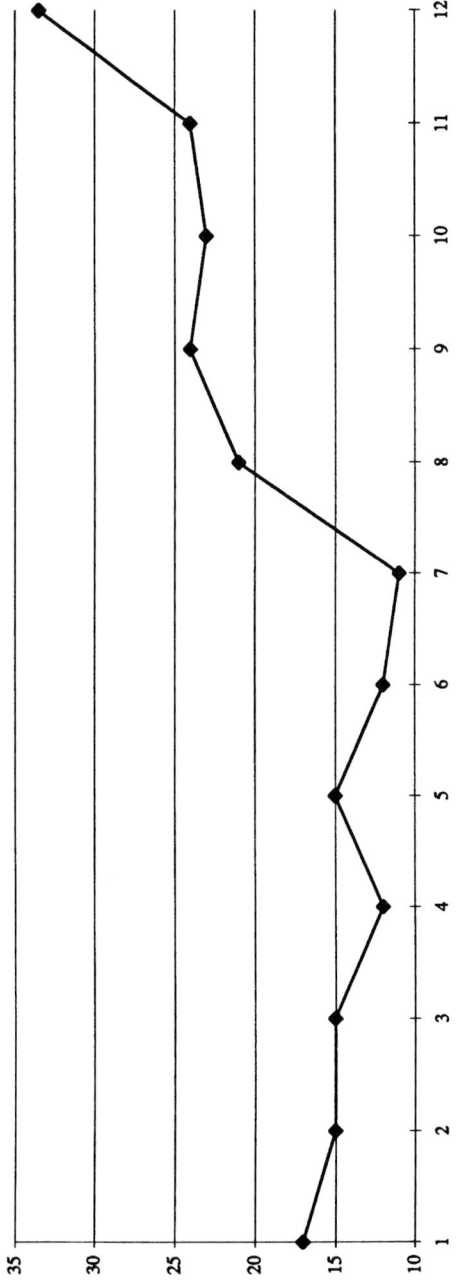

Figure 10.2   Repetition rates for Africans (source: RIEP, 1991, *Education and Manpower Development*).

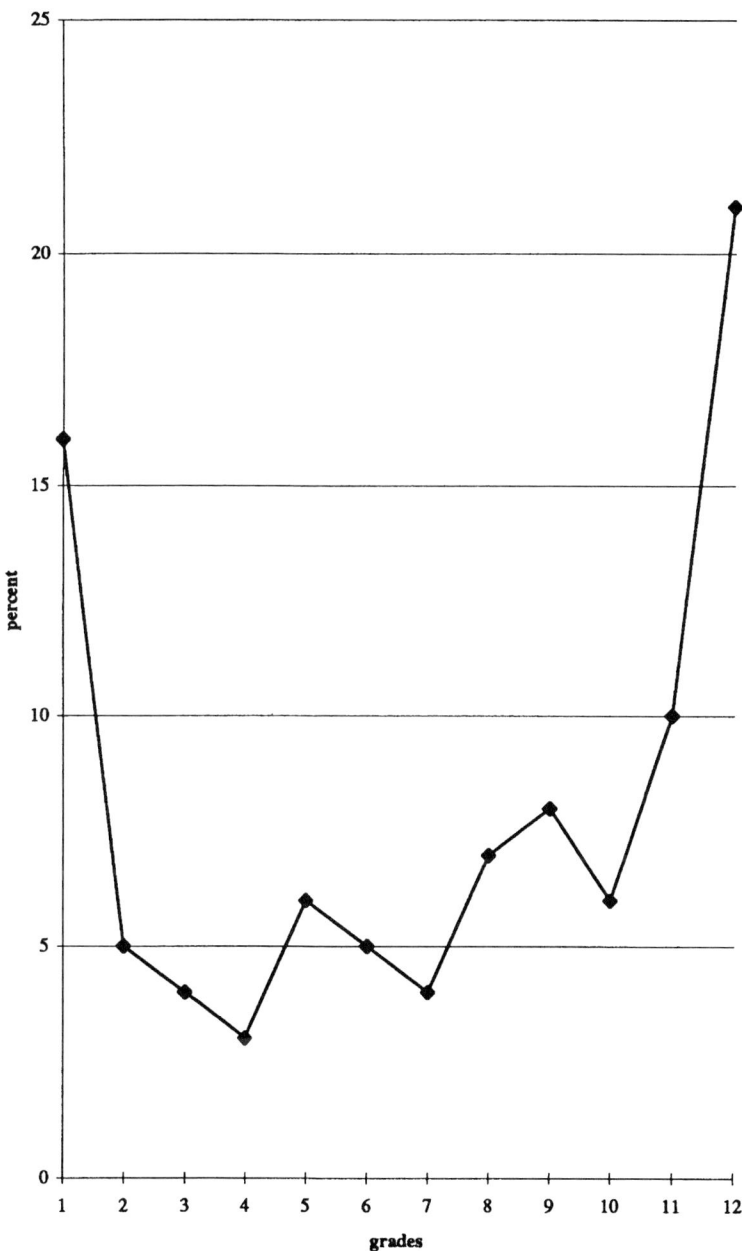

Figure 10.3 Estimated drop-out rates for Africans (source: RIEP, 1991, *Education and Manpower Development*).

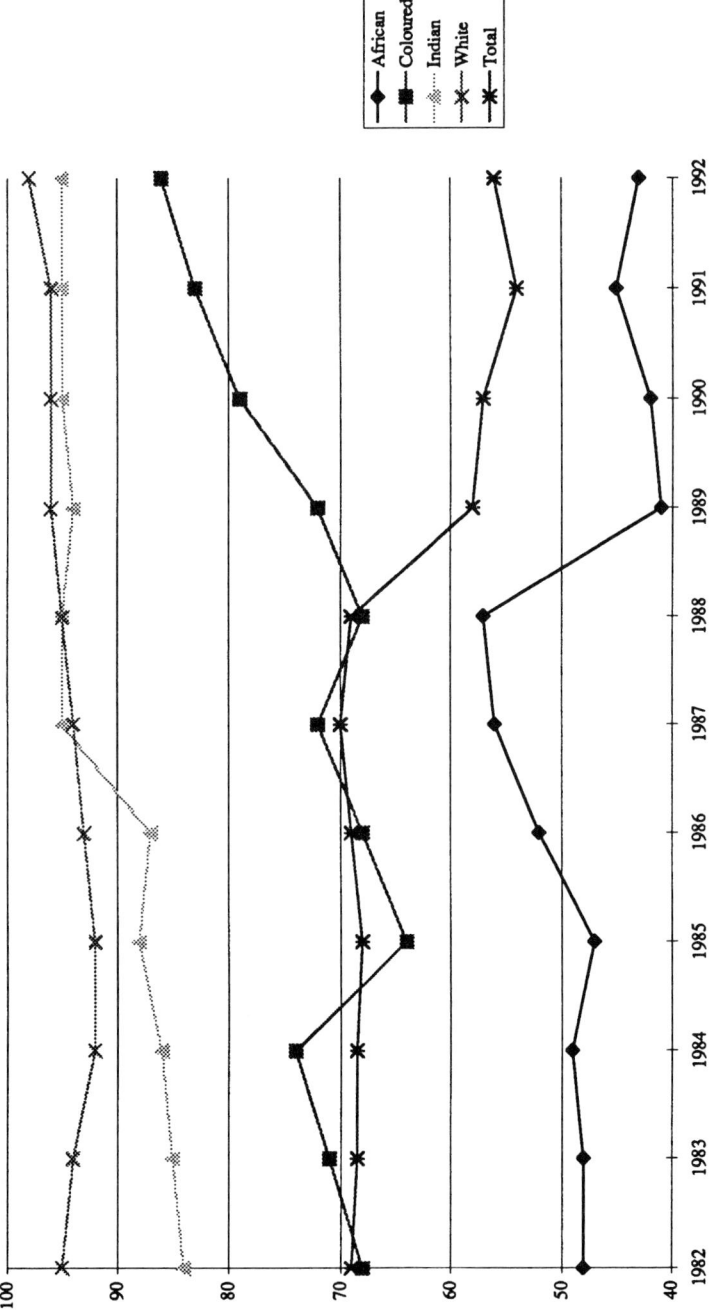

Figure 10.4 Matric pass rate 1982–92 (source: SAIRR, *Fast Facts*, 1993).

Boards (Edusource, 1994). In 1993 the pass rate fell to just 38% with only 7% satisfying university entrance requirements. These results are in stark contrast with those of the other departments where pass rates have consistently averaged about 93% (98% for whites, 96% for Indians, and 86% for coloureds (SAIRR, 1993).

Malpractice in the examination system

A close study of the DET examination process reveals many disastrous practices in this examination. There is a strong belief that the poor African pass rate was a purposeful strategy of the state to keep blacks perpetually in a position of disadvantage (Dlamini, 1991). This is seen as being orchestrated through the strategies of carelessness that surround the work of the DET; the authoritative and secretive processes of the same department; the lack of resources put into DET classrooms; and a political commitment to punish those who oppose the government and reward its allies. In response to this perceived vendetta, a vicious circle of anger and frustration resulting in boycotts and poor results was set in motion. The examinations in particular have come to be a flash-point in this cycle. In addition, the statistical adjustments which are made to marks during the awarding process are not widely known and are perceived to be shrouded in mystery. These, coupled with anecdotal evidence of gross misconduct in examinations, has led to the public having little faith in the largest public examination system in the country.

Examinations and the qualification system

Examinations in South Africa act as instruments of selection and are enjoying a pre-eminent status as they lead to certification, and certification has an extremely high currency. In a country where certification enjoys such an important position, qualifications become the barriers to progress along various education and training career paths. In this way, then, examinations dominate the entire process as students compete for the fewer and fewer spaces in the labour market and tertiary institutions.

It has been extremely hard for the majority of students to complete high-school education. They leave formal education with no qualifications or means of access to anything except unskilled labour, when it is available. The current qualification structure is embedded in the school system and alternative routes to qualification are not helpful. There are some education programmes aimed at adults, but these are inadequate and do not meet the qualification needs of adults, especially the unemployed and those from rural areas. Conversely, some forms of alternate provision falling outside the formal schooling system have not received recognition from the higher education institutions. In addition, we face a situation where there are huge numbers of children out of school and many young people described as 'marginalized'. This, coupled with rural poverty and the mushrooming of informal settlements near to urban centres, makes the need for education and training programmes that are linked to comprehensive human resources strategies of vital and urgent importance.

South African industries conduct a significant amount of trade-related training. This was necessary since the school curriculum was largely academic and those schools which did offer vocational education were outdated and/or underresourced. Currently many organizations offer training, but the accreditation of centres for the purposes of training and testing is not centrally organized. For example, the numerous training boards do their own training and testing. The qualifications gained are industry-specific and have no transferable value.[2] The other avenue available is work-based training, access to which has in the past been tightly controlled by white trade unions.

There are problems too with the certification process. The SAFCERT was established specifically to issue certificates and establish equivalences across the different examination boards. Because education delivery and examinations are heavily embedded in the school curriculum, this does not accommodate education and training taking place outside the formal system. The central control of certification by SAFCERT has also perpetuated the disempowerment of important stakeholders in education, including the examination boards. Because of heavy control of the certification process, the examination boards cannot give credit to the product they have assessed. Employers and other important stakeholders do not have a say in the qualification-making system. Overall, SAFCERT perpetuates the policy of 'equal but separate', and hides the real issues by drawing attention away from the reality of differences in standards in education caused by a fragmented and unequal system.

Proposed changes

From the end of 1994 education for all South African children is being provided under one ministry, and by the year 2000 the process of transforming the education and training process will be in place. Transforming formal education will include, among other things, introduction of 10-years compulsory general education for all South African children. It is proposed that this will culminate in an externally examined qualification, the General Education Certificate (GEC). The GEC will be an important qualification in the country. For many it will be an exit certificate attesting to their achievement during compulsory education. For others it will be a stepping stone to further studies either in schools or through a work-based route.

Alongside the formal education system there will be provision for adult education that will also lead to a GEC equivalent to the formal education certificate. Beyond this point, there will be an open learning system that will be accessed by learners via different forms of provision as Fig. 10.5 demonstrates. For those in formal schooling, it will take three years to complete this part of education and the exit certification will be known as the Further Education Certificate (FEC), acquired at the age of 18/19. Those outside the formal system will be able to accumulate credits toward the FEC through various routes such as community colleges, adult basic education programmes, distance learning, work-based training, trade training, Public Works programmes, etc. Beyond this point all learners should have an equal chance of accessing further education, training and employment. Figure 10.5 illustrates how the system will work.

The above national qualification structure will enable qualifications, learning and training acquired from elsewhere to be recognized formally. This will make the integration of education and training possible as emphasis will not be on academic education only. This system will also make it possible for learners to access education and training as a lifelong process. The main thrust of the new system is that education should no

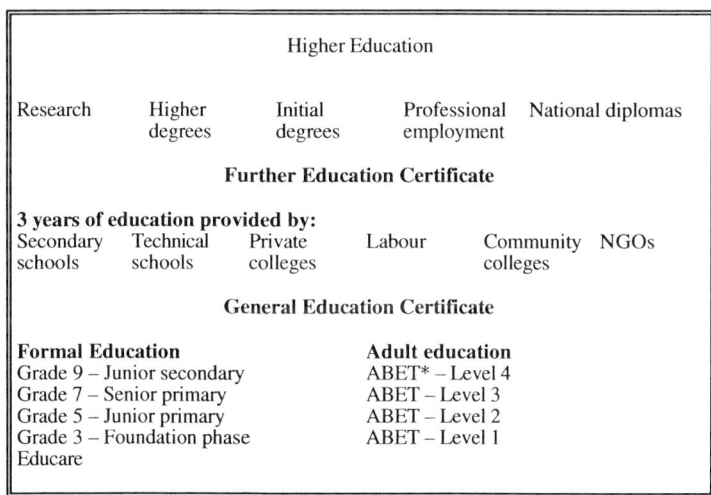

Figure 10.5 National qualification framework (source: South African Qualification Authority Preparatory documents, 1994). ABET: adult basic education and training.

longer be fragmented racially, the separation of academic and vocational training should cease, and a coherent structure of delivery will be in place with identifiable paths of learning that are easy to access.

Implications for the future system

For many years in South Africa we have postponed the impact of a public examination for up to 12 years of education, and the introduction of an external examination at an earlier stage might seem a retrogressive step. The previous experiences of external examinations before the end of high school, at the end of primary school and junior secondary (Standards 6 and 8), demonstrate that they were definitely exit certificates for many students for a long time. It is interesting to note that these examinations continued longer for the African students than for the white students and contributed significantly to the low participation of the former in education. There is a possibility that the GEC may cause a bottle-neck at this point and will then have to be looked into carefully.

The real worth of a qualification will be determined by the opportunities that are available for further education, training and work. However, educational systems are generally vulnerable to the economic circumstances in which they operate, including how the human resource development systems are structured. It can be postulated that one of the reasons for phasing out the qualifications below the end of high school was the non-availability of opportunities for training and work opportunities after this qualification[3] as the minimum requirements for training and work increased. As yet, there is not enough information about the real opportunities that will be available beyond the basic compulsory education phase. Will education be financially inaccessible to many beyond the GEC? What will be the real opportunities for those who have to leave the system? What is expected by the end-users of this qualification?

The establishment of a single Ministry of Education, with nine provinces, will bring about an end to divided control in education, but a different kind of control will emerge as the provincial departments get established. The differentiated education system of the past has resulted in differentiated curriculum delivery in the various school systems. The differences in curriculum delivery have resulted in unequal quality of outputs as measured by student achievement levels for example. But there is a dearth of systematic and usable information on the performance of the various education systems prior to the introduction of the new system. So far, we have information that compares educational outputs across the previously segregated departments only at Standard 10, or end of high school. The results[4] at this level show a definite correlation of failure with race in South

African education (National Education Policy Investigation, 1992). This perception is easily generalized and has become a common assumption even at the lower levels of African education. Information about the achievement levels of all students at all levels of education is needed to understand the extent of differences, and to plan appropriate interventions.

The use of external assessment to set an exit standard for compulsory education will substantially alter the shape of the provision up to that point. This examination is likely to affect the curriculum even at the lower primary phase, influencing the nature of subjects to be studied, the learning outcomes to be promoted and also the nature of assessment arrangements that are made. At present, our schooling system is designed around seven years of primary education and five years of secondary education. The proposed qualification structure, together with the national education policies to make education compulsory up to the end of Grade 9, creates a new education structure having only three years at the senior secondary level, and brings an important qualification level down. Schools are not structured in this way at present, and the pressure brought to bear by an external examination will be felt even earlier in the system.

South Africa is a linguistically heterogeneous country. This arises from the fact that the majority, Africans, have command of many African languages but may not have full command of English or Afrikaans. On the other hand, the minorities of the country have full command only of English or Afrikaans or both. This situation has influenced the emerging national policy on language that portrays English as a common language for all (DNE, 1994). This obscures the fact that the majority of African learners may not have full access to the language, especially at the lower levels of education. The advisability of singling out English as an examination and only language in the core curriculum for students who have poor proficiency in the language and who rarely if ever use the language outside school has to be perceived and addressed as a problem. Students with limited knowledge of a language will inevitably be handicapped in the acquisition of knowledge and skills presented in a non-native language as well as in their ability to demonstrate in examinations the knowledge and skills they have acquired. It is important to understand the weaknesses and strengths presented by students' language acquisition levels and how these will impact on a certification process at this level.

The relationship between the inputs and outputs or achievement levels in education is generally accepted as fundamental by those who are concerned with the quality of education. Most studies, though, have focused narrowly on 'basic skills achievement' that is measured by standardized tests as sole indicators of outcomes in effective schools (Jansen, 1994). This could be expected as researchers interested in this area tend to have a sociological interest in education and are not assessment specialists. The emerging interest in South African research in the quality of education (inputs and outputs)[5] comes at a time when education provided for the majority has degenerated to unacceptable standards. More than ever, we need to have baseline information on the nature of the inputs to our schools. However, we need to avoid the pitfalls experienced in other countries by using qualitative indicators of our outputs if we are to improve the quality of education through control of our examinations. This period of change presents an opportunity to evaluate systematically both inputs and outputs and measure progress over time.

A qualification structure which assumes that all learners can be assessed in the same way can contribute to low achievement rates. In other parts of the world, it has long been realized that academic education is not relevant to all learners and therefore curricula tend to reflect divisions along the lines of those who will pursue academic vocations, and those who will get into industrial and commercial vocations. The potential problem with the proposed integrated education system is that it could encourage dualism in education if the public does not believe that the vocational route will allow learners to proceed to the highest levels. This is likely if the vocational options are grafted on to the existing old general education programme. It is therefore important in planning for the new system that vocational and general skills are taken as equally important starting points in curriculum development. It is equally important that vocational curricula should not ignore theoretical work or the key 'domains' promoted in general education. But more than ever it is the assessment methods that will produce an effective qualification system and that is a big challenge.

It is therefore the extent to which both vertical and horizontal articulation can be shown to work that will inspire the confidence of stakeholders. In other words, it is when the system can allow for easy movement across the different modules, as well as through the modules, to institutions of higher education that the integrated path will have credibility. Assessment therefore should be based on establishing linkage between the general and vocational routes. For example, every practice has a theory behind it and this provides a common background for assessment across a diversified path. Common competencies can sometimes be found to link areas which may appear to be distinct. Defining the expected outcomes at a particular level will provide a formula for equivalency across the different provisions. This is no small task that lies ahead.

Conclusion

South African education has in the past been characterized by high standards for those receiving privileged education. Unfortunately, the same cannot be said for the majority of learners. The system that catered for the latter has been characterized by very low achievement and progression rates. They have not only suffered more from the racially fragmented system, but have benefited least from a punitive assessment and examination system.

One of the key problems of the South African education is the lack of a coherent structure in the qualification and certification process. The proposed national qualification structure is aimed at addressing this problem. What is more important about the future education and training system is that for the first time, learners may gain many chances to be assessed and gain qualifications through different routes.

The concept of lifelong education is long overdue as there are already many casualties of Apartheid education. However, these changes are accompanied by many problems, challenges and opportunities. The introduction of an examination at an early stage might create problems, especially if it is unrelated to a comprehensive human resource development plan. The regionalization of education will create different problems for examining bodies. How do we know if all assessment practices are of equal standards? The fashionable rhetoric is of learning outcomes that will make articulation possible across all provisions. But how does one operationalize this in assessment? Above all, we need to know whether all learners are achieving desired outcomes, and, if not, to have a way of influencing policy in this direction. In the past, assessment has not been considered to be part of the process of delivering education, except as a final, summative process. Our assessment practices have been primarily geared toward selection, and we do not have experience in using assessment for monitoring and supporting learning and teaching. South Africa must quickly catch up with the latest developments in this area if it is to improve the quality of educational delivery. The implementation period will be more challenging than the policy-making phase from which we are emerging.

Notes

1. The 'progression rate' is defined as the proportion of a cohort or year-group which moves to the next grade or phase at the end of a school year.
2. Under the old system, nineteen major industries had their own schedule of training. This led to a situation where, for example, separate training schedules existed for a fitter and a turner in the mining industry, a fitter and a turner in the explosives industry, a fitter and a turner in the railway industry, and so on.
3. Minimum requirements for many training programmes like teaching, nursing, policing, and trades training used to be at the Standard 6, then at Standard 8 levels. Most jobs could be accessed with a Standard 6 and then Standard 8 certificate (e.g. public works, transportation, post office, etc.)

4. For example the DET Standard 10 pass rate has hardly surpassed the 35% level for the last two years, while the white students' pass rate has constantly remained around the 96% level.
5. See, for example, the research on understanding and improving education quality in South African schools and classrooms carried out by research units like the Policy Support Unit of Education Foundation, the Education Policy and Systems Change Unit (EDUPOL) of the Urban Foundation, the Foundation for Research and Development, Education Policy Units educational Departments of universities, and the Institute for Improving Education Quality (IEQ).

References

DNE (Department of National Education) (1994). *South Africa's New Language Policy*. Discussion Document. Johannesburg: DNE.

Dlamini, T. C. (1991). *A Study of Some Aspects of African Matriculation Examinations*. University of Natal, Education Projects Unit Publications.

Edusource (1994). *Data News*, June 1994.

Jansen, J. (1994). Beyond effective schools. Paper presented at IEQ Conference on Effective Schools, Cape Town, March.

National Education Policy Investigation (1992). *Curriculum*. Oxford: Oxford University Press.

South African Qualification Authority (1994). *A National Qualification Framework: Implementation Plan, Including the Establishment of the SAQA*. Draft Document. Johannesburg: Centre for Educational Policy Development.

RIEP (Research Institute for Educational Planning) (1992). *Education and Manpower Development*. Bloemfontein: RIEP.

SAIRR (South African Institute of Race Relations) (1993). *Fast Facts*. Johannesburg: SAIRR.

CHAPTER ELEVEN

Quality improvement through testing: Egypt's reform programme

Fouad Abu-Hatab and David Carroll

Background

Traditional methods of oral, written and practical examination, based on learned tasks, have been used for thousands of years in Egyptian education. Educational assessment was formally established in the first Islamic University in Egypt, Al-Azhar University, and its satellite multilevel, one-teacher schools called 'kuttabs'. Evaluation in this type of education was in terms of Islamic educational objectives. For example, memorization of the Quran was evaluated by both oral and written examinations. Modern education, following the French system, was introduced in the first decade of the nineteenth century by Mohammed Ali, founder of modern Egypt. New primary and secondary schools and higher education institutes were established. When Egypt became a British colony in 1881, the English system of education dominated. These two parallel systems, of religious Islamic and secular education, coexisted, and have been in conflict ever since.

Figure 11.1 shows the basic structure of Egyptian education in 1908, immediately prior to the creation of the first National University (King Fouad I University, later Cairo University). For approximately 50 years thereafter, this basic structure survived, with variations. For example, after World War II, an Education Act similar to the English 1944 Act was introduced. A form of 'O level' (at the end of the 4th Grade in secondary school) and 'A level' (at the end of the 5th Grade) was adopted. In addition, two different tracks of secondary education developed, general academic (equivalent to the English grammar schools) and technical/vocational. However, education was expensive, and except for a free, compulsory closed-ended education lasting four years, fees were charged at all stages.

This period came to an end following the 1952 Revolution, which established free education in all state educational institutions at all stages as a

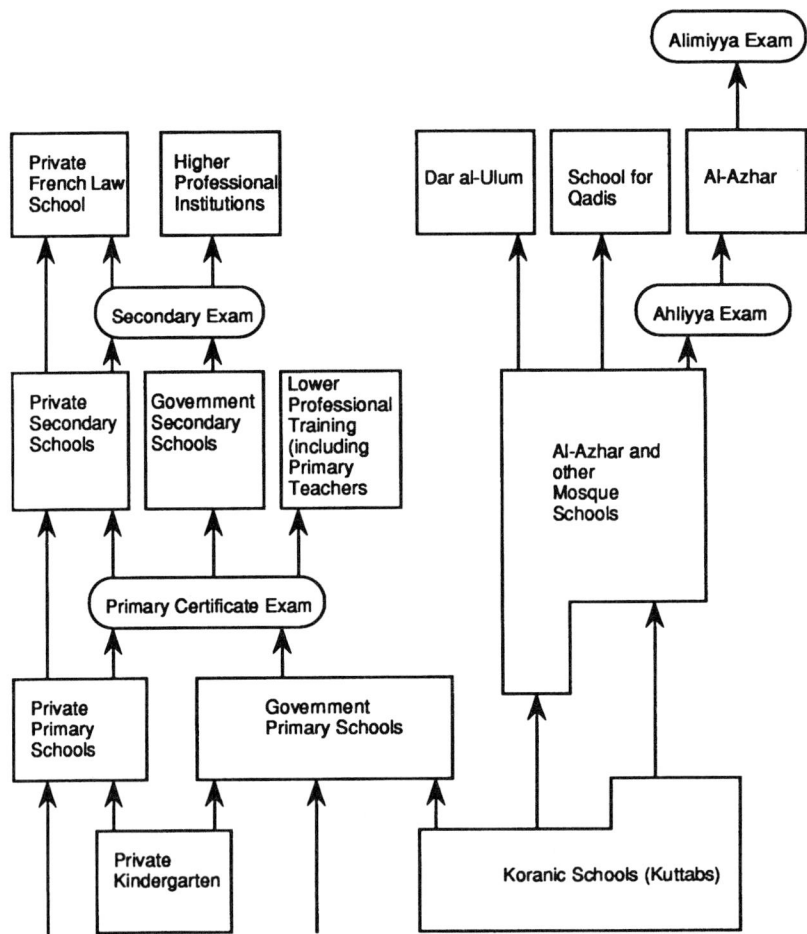

Figure 11.1   Egyptian school systems in 1908 (source: adapted from Reid, 1990: 16).

right guaranteed by the Constitution. This led to a dramatic expansion in all phases of education. Demand for university education was particularly strong, partly as a result of the policy which guaranteed positions in government service to university graduates. During this period, the O level and A level were abolished, and preuniversity education was divided into three stages, with corresponding national certificates:

1. Primary school certificate after six years of compulsory general education (students aged 6–12 years)

2. Preparatory school certificate after three years of intermediate education (students aged 12–15 years); and
3. Secondary school certificate after three years of secondary education (students aged 15–18 years).

These national certificates became the effective criterion of 'ability to benefit' from the succeeding stage, a situation that has persisted to the present. A pass in the preparatory school certificate is required for admission to general secondary education, and the aggregate score in the general (i.e. not technical/vocational) secondary school certificate (GSSC) determines not only admission to university, but which faculty a student can enter. A high aggregate gains admission to a desirable faculty such as medicine or engineering; a low aggregate condemns a student to the faculty of education, or a four-year 'institute'.

The three-stage system remained unchanged until Education Law No. 139 of 1981 extended the period of obligatory education to nine years, and combined primary and preparatory school in one entity known as 'basic education'. The primary school certificate became in effect a promotion test, and certificates were awarded only at the end of Grade 9 (Preparatory or Basic Education Certificate) and Grade 12 (the end of secondary schooling).

Continuing policy concerns

There is a number of recurring themes which characterize debate about assessment amongst policy makers in Egypt:

Rote memorization

The level of rote memorization in the examinations is a concern because it is accepted that excessive memorization does not help students develop life skills. Memorization also encourages private lessons and cheating in examinations, and higher-skills questions are therefore seen as a way to strengthen free education, by making cheating more difficult; but there is no consensus about how to introduce higher-skills questions.

Drop-out and grade repetition

Rates of repetition and drop-out are high. The average child takes more than six years to complete a five-year primary course. Formal promotion testing begins in Grade 4, so repetition is concentrated in Grades 4 and 5, and the typical primary drop-out has 4.5 years of schooling. The examination system contributes to this by failing to give adequate information about learning, and not specifying the remedial action required to avoid repetition.

Admissions policy for university

The constitutional entitlement to education, coupled with a 'graduates' policy' entitling all university graduates to a government post, has created a dilemma no minister has succeeded in resolving. On the one hand, the difficulty of finding enough government posts has increased to the point where graduates have to wait six years for a government post. The bureaucracy has become bloated and inefficient, and bureaucrats' real salaries have declined substantially. The quality of university education is generally poor because of the scale of the expansion of numbers and the impossibility of funding it adequately. On the other hand, the education system has played a crucial role in creating Egypt's huge first- and second-generation middle class, and giving them a stake in continuity; but this, by helping a much larger proportion of the population achieve a better standard of living and education, has also increased the number of people with fears about the future. Therefore, the minister is faced with three unpalatable options: allow the situation to continue to deteriorate; reduce numbers admitted to university in line with employment opportunities; or, delink higher education from employment.

University admission is effectively only available to those who take the GSSC, thus disadvantaging graduates of technical/vocational secondary. This was tolerable when fewer than 20% of students were in technical/vocational secondary, but introduces a major distortion now that proportion is approaching 70%. This is acknowledged; but no reform proposals have been produced.

Certification

There is growing recognition of the need to provide certification for those leaving formal education. The present examination system does not do this for most leavers, because a pass at each stage entitles the student to continue to the next stage. Therefore, leavers tend to have at most a certificate of completion, and employers tend to take little account of educational qualifications. This forces students to stay in education for as long as possible, rather than seeking a specific qualification.

Popular concerns

There is also a number of recurring themes which characterize popular debate about assessment:

'Mediation' in testing

Promotion depends on passing tests. Teachers who control the tests can use this power to 'encourage' students to take private lessons and use external books. Therefore, while the examination system is not itself a cause of private tutoring, fear of such 'mediation' forces ministers to take measures which may be educationally undesirable, such as cancelling continuous assessment, in order to maintain fairness.

'Group cheating'

Some teachers illegally help their students to pass external examinations. There was a rash of such cases in 1987, to which the then minister responded firmly, fining a local director of education, suspending teachers and cancelling the results of a total of 609 students. He had broad popular and political support in this action, and since that time, the problem has been much less evident.

Increasing access to higher education

In spite of the constitutional guarantee, demand for admission to university far exceeds supply. Citizens therefore seek to ensure that the government does not attempt to limit the number of passing students by making the examination more difficult. Questions are studied by the press and public to assess whether they are within the ability of the average student, and the ministry 'calibrates' grading schemes by test-marking a sample of papers. Where a question is found to be 'too difficult', its mark allocation is reduced, and the marks redistributed. Any innovation, such as the introduction of questions testing higher-order skills, is suspect for the same reason, and there is popular demand for an increase in the number of times students can sit the examination, to improve their grades.

Controlling 'back door' access to higher education

Fear of 'mediation' in university admission has led to a visibly impartial system of 'coordination' of admission by computer. There are however legitimate ways to bypass this. IGCSE allows those who can afford an English-medium education to enter university with seven minimum passing grades, and gives preferential access to high-prestige English-medium faculties. This is resented by some, and an unsuccessful effort was made in the late 1980s to eliminate foreign examinations. Students can also transfer from foreign universities (which do not require GSSC) into the national university system.

Beirut University has a branch in Alexandria, and Khartoum University a branch in Cairo, which students who can afford the fees use to get into courses in the national universities for which their GSSC grades would not have been adequate.

However, the biggest such scandal of recent years was the result of an attempt to increase access by waiving normal entry requirements for mature students, who would pay fees and attend part time. In practice, universities enrolled many new or recent secondary graduates, who were able by paying to enter a better faculty than their grades would qualify them for. The government was therefore obliged to establish a mandatory five-year break between leaving secondary education and applying for this type of university education.

Recent reform efforts

Educational reform became a national priority in Egypt from the mid-1980s on. One great reformer of this period was H. E. Dr Ahmad Fathi Sorour, Minister of Education from 1986–90. His multifaceted strategy for educational reform is described in *Towards Educational Reform in Egypt: A Strategy for Reform and Examples of Implementation* (Sorour, 1990). One particularly controversial innovation was the reduction of basic education from nine years to eight. Although it was supported by the results of an evaluative study of an experimental eight-year school established in the early 1980s and funded by Germany, this decision was based essentially on economic necessity. Particularly controversial was the implementation of this decision by combining the 1988–89 grades five and six into a double cohort. This double cohort (spread over three years by grade repetition) will increase the number entering the GSSC examination from about 260,000 in 1994 to about 400,000 in 1995 and only slightly less in the two following years.

Since 1989, when Education Law No. 233 of 1988 came into force, the formal education system in Egypt has consisted of two levels:

- basic education for students aged 6–14 years, divided into primary (five grades) and preparatory (three grades)
- secondary education for students aged 14–17 years (three grades).

Figure 11.2 maps the Egyptian education system current in 1993–94.

The principles for examination and certification remain unchanged for both levels. The Ministry of Education, centrally and locally, is responsible for administration of examinations. The GSSC is organized at national level, by a directorate of the Ministry of Education. The same examination is taken nationwide. The Basic Education Certificate is organized by the

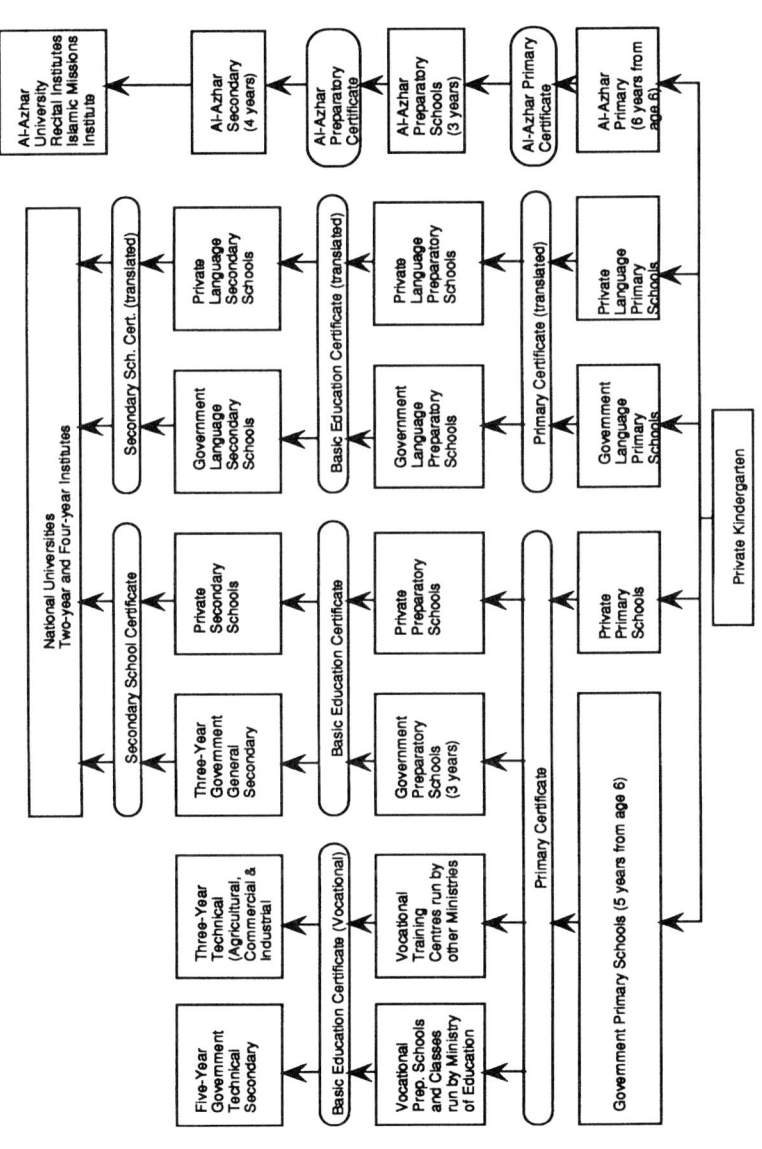

Figure 11.2 Egyptian education 1993–4.

governorates (the 26 major administrative regions into which Egypt is divided). A different examination is taken in each governorate. Other examinations are organized either by the educational subdirectorate (idara) or locally. The content of all examinations is controlled by centrally-prepared specifications. The GSSC papers are prepared by committees formed by the central ministry. Similar committees are formed at governorate level for the Basic Education Certificate. Teachers design the annual promotion tests.

A 1981 conference in the Ministry of Education on the theme *Modernising evaluation is the pathway to educational reform* led to the establishment of the Supreme Council for Examinations and Educational Evaluation, which was responsible for monitoring general certificate examinations. This council initiated publication of a *Model Question Book* for each subject, to facilitate assessment activity for both teachers and students. The British Overseas Development Administration also provided aid for the establishment of an item bank.

In June 1990, a Centre for Development of Examinations and Educational Evaluation was established by ministerial decree. A few months later, in November 1990, this centre was transformed by presidential decree into a national centre, the National Centre for Examinations and Educational Evaluation (NCEEE). A Centre for Curriculum and Instructional Materials Development (CCIMD) was also established in 1990. Both centres were offshoots of the National Centre for Educational Research and Development (NCERD), first established in 1972.

The NCEEE has five departments, as described below.

Test development
• Responsible for examination specifications for all stages, levels and subjects, producing items for item banks, especially for general certificate grades 3, 5, 8, 11, and authoring student self-evaluation guides in various subjects, primarily as a supplement to student textbooks, but also to assist teachers and parents.

Research
• Responsible for carrying out research in evaluation. The most recent one is the 'Grade 4 experiment', whereby the centre took over all aspects of examination development and administration for the promotion test at the end of Grade 4 within two governorates. This was also both an internal evaluation of NCEEE procedures and an external evaluation of the ministry's curriculum, teachers, school administration, facilities, buildings, etc. This was an opportunity to develop and field test systems and procedures on a manageable scale. Further activities of this department include research on psychological testing and statistical analysis relevant to testing problems.

Training and dissemination
• Responsible for development of a comprehensive training plan, training trainers for many programmes and development of skills in writing high-level cognitive items, observation checklists, etc.

Operations and information
• Responsible for problems of item banking, a database for examinations information, computerized testing and a computer-managed assessment programme.

Evaluation department
• Responsible for monitoring the education system as a whole. The experience of the inspectorate system in Britain is guiding the work of this department.

Progress of the reform

From its inception, the NCEEE was responsible both for examinations at all stages, and for evaluation of curriculum, teachers, systems and institutions. Its first task was reform of the GSSC examination. For most of the 1980s, policy was to reduce the number of students in higher education. Numbers entering universities were reduced from 93,486 in 1983–84 to 69,949 in 1990–91. Since the sole criterion for admission to university is the aggregate score on the GSSC, general secondary education is commonly perceived primarily as a filter for university admission, rather than as an end in itself. At the same time, reform of general secondary education was severely constrained. Because of its central role in the university selection process, any change in the examination was difficult and controversial.

Therefore, reform had two goals—developing the certification function of the GSSC, and hence the value of secondary education as an end in itself; and making selection for higher education as valid and fair as possible, in view of the increased competition. So long as success in the GSSC entitled a student to a place in higher education, very few GSSC graduates would enter the labour market. Therefore it was proposed that a GSSC pass should be a necessary, but not sufficient, criterion for university admission. This would allow a larger proportion of students to achieve certification. Since the examination would have lower stakes, the content could become more representative of the skills of the average student, and changes in the type of skills assessed would provoke less resistance.

At the same time, an advanced-level examination was proposed, in which students would choose at least two qualifying courses relevant to the specialization they wished to study. Admission to higher education would be based on both the GSSC and the qualifying A-level courses together with other assessments—e.g. aptitude tests for music, art, physical education, etc.—where appropriate. In this way, the predictive validity of the selection process could be increased.

This policy was welcomed by experts from both USAID and British ODA, but before it could be introduced, there was in 1992 a reassessment of priorities, and education was redefined as an issue of national security, both economic and social—'Education problems are problems which affect the Egyptian family as a whole and affect the national security of Egypt'.[1] Given the public anxiety about the proposed changes to the GSSC, and the hardship the people of Egypt were already undergoing as a result of structural adjustment, a decision was taken not to subject them to additional hardship from reform of the GSSC.

Numbers entering university therefore increased substantially from 1992–93 on.

> [In 1992] larger numbers of students were accepted into the universities than ever before. Our objective, as instructed by President Mubarak, is to create more opportunities for young people to finish their university education, thereby fulfilling the Egyptian family's ambition to provide their children with a higher education.[2]

Although at the time of writing definitive figures for 1994–95 are not yet available, the number admitted was about 120,000. In 1995–96 and the two subsequent years, it may be as high as 200,000, to accommodate the double cohort.

At the same time, measures were taken to reduce the stress caused by the GSSC:

- From 1995, the examination will be taken over two years (the second and third years of secondary education), instead of one.
- Students who fail or get a poor grade on the first stage will be given the opportunity to resit the examinations in August, and improve their total grades.
- Students are allowed to take the examination up to four times. This will reassure them, because they know they can have another chance, whereas with only one chance, if they waste it, their future is damaged beyond repair.
- Optional subjects have been introduced, giving students some opportunity to choose subjects suited to their interests, inclinations and potentials.

Major change in the content or skills tested in GSSC has been put off indefinitely. Instead the primary focus of policy since 1991 has been basic education. Following the 1990 Jomtien conference, President Mubarak declared education the national project until the year 2000; and Egypt has been selected as one of the nine countries which will address issues of basic education for the rest of the decade. Commitment on the national level to the EFA declaration was reaffirmed at the ministerial meeting in Paris in June 1993.

As part of this national project, a Conference for the Development of Primary Education Curricula was held in Cairo in February 1993. This resulted in a new curriculum to be introduced starting in 1994–95, dividing the five-year cycle of primary education into two levels:

- Grades 1–3 with an emphasis on Arabic language skills, religious education and mathematics, supplemented by a broad course of activities and practical skills in daily living and community life, including primary forms of science, social studies, environmental studies, practical manual skills, music, art and physical education.
- Grades 4–5 with a continued emphasis on language, religious education and mathematics, and the introduction of separate subjects, for example science, social studies, music, art, physical education and practical skills.

The major recommendations of the conference with regard to primary school assessment were as follows:

- basing evaluation on mastery learning, emphasizing diagnosis and remedial testing
- comprehensive evaluation of the primary school experience, including written, oral, practical and performance tests
- re-emphasizing educational activities neglected by examinations
- general assessment at the end of grade three to ensure that pupils have acquired the basic skills of reading, writing and mathematics, and of grade five to assess reading, writing, mathematics, science and social studies.

A ministerial decree was issued in March 1993 establishing two-stage primary education and general examinations at the end of Grades 3 and 5, and giving the NCEEE responsibility for the system of examinations, in cooperation with the Ministry of Education and the NCERD.

The main aim of the new examinations, to ensure that all pupils acquire basic skills in Grade 3, and a higher level of proficiency in Grade 5, could not be achieved by the present localized examination system at these grades because:

- there is no external monitoring of standards; the only intervention consists of the publication of non-mandatory examination specifications
- the examinations produced locally do not focus primarily on priority skills, and question writers have not been adequately trained.

The new assessment aims to set a standard which all schools should meet in due course, thereby ensuring that participating students achieve an acceptable basic education, and helping to restore parents' and students' faith in the benefits of that education. In addition, the assessment aims to develop the diagnostic function of evaluation by linking the tests to remedial activities, and offering individuals 'at risk' a second chance to take the examination after remedial work.

However, the NCEEE does not have sole responsibility for the assessment. Governorates have a high degree of administrative and budgetary independence, and each governorate has its own examination. The original plan for the assessment divided functions between the NCEEE and the governorates as follows:

- The NCEEE produces a national test specification taking into account governorates' views.
- The NCEEE writes, reviews and field-tests test questions.
- Each governorate's test is assembled jointly by NCEEE and governorate staff.
- Camera-ready copy prepared by the NCEEE is supplied to the governorate.
- Each Governorate prints, administers and grades its own test.
- Monitoring and research analysis is undertaken by the NCEEE.

The first round of the assessment did not go according to plan. In effect, the governorates refused to use centrally-prepared items, and the Grade 3 and 5 examinations in 1994 contained no NCEEE test items.

There has been considerable debate about how the assessment can be put back on track. It is reasonable to speculate that the Governorates' main objection was to increased central intervention, and to the possibility that ultimately the ministry would be able to monitor standards effectively. Therefore, plans for the 1994–95 round of testing were restructured so as hopefully to gain the cooperation of the governorates by four types of measure:

- increased consultation
- improved specification design and dissemination
- training and other services to mudiriat
- emphasis on development of remedial work.

For the foreseeable future, use of NCEEE items in governorates' tests will be voluntary.

The system as originally proposed would have achieved significant improvements in test quality for very small outlay, and established consistent standards and criteria for success nationwide. In the event, however, it could not be implemented, and major rethinking was required. The consistency of standards and high-quality test material that would have been achieved by the use of centrally-produced precalibrated items has temporarily been sacrificed, as has the possibility of making comparisons between governorates, or schools in different governorates using tests of known difficulty.

However, even this reduced plan will lead to some benefits. The development of more rigorous examination specifications and monitoring of their implementation will lead to some convergence of standards, and provide a tool for improving the quality of question writing. The emphasis on development of remedial classes during the summer break will provide a means for reducing grade repetition without relaxing standards. The new channels for consultation, and the emphasis on offering training and other services to governorates will both strengthen local capacity and build a basis of trust between the governorates and the NCEEE. However, this is not of itself enough to achieve the original aims. At the time of writing, the future of the Grade 3 and 3 assessments is still very much in the balance.

After a long period of slow change, Egypt has, since 1985, been going through an educational revolution. Under Dr Sorour, the pace of this revolution was hectic, and it would have culminated in 1995 with the reform of the GSSC examination; but national and international events have led both to a change of emphasis and the adoption of a gradualist approach.

Notes

1. From an interview with H.E. the Minister of Education, Dr Hussein Kamel Bahaa El-Din published in *Al-Akhbar* newspaper on September 7 1992, (by Mahmoud Arif).
2. Ibid.

References

Reid, D. M. (1990). *Cairo University and the Making of Modern Egypt*. Cambridge: Cambridge University Press.

Sorour, A.F. (1990). *Towards Educational Reform in Egypt: A Strategy for Reform and Examples of Implementation 1987–90*. Kalyoub, Egypt: Al-Ahram Commercial Press.

CHAPTER TWELVE

External assessment in Korea

Chan-Jong Kim

Introduction

The government of the Republic of Korea is highly centralized. The Ministry of Education has strong control over major educational policies. A series of five-year economic plans since 1962 have yielded strong economic growth (Son, 1987). Many people believe that education has been one of the major factors that have contributed to the national economic development. At the same time education has been an area with many controversies and problems yet to be solved.

There have been continual reform movements in education, and one of the major targets was the entrance examination system at various levels. A brief introduction about the educational system and the social and educational contexts in Korea is provided. Major changes and their backgrounds of external assessment since the 1960s are also discussed.

The Structure of education in Korea

The Education Law promulgated in 1949 declares the adoption of a school ladder following a singular track of 6-3-3-4. An overview of the structure of education and the approximate age of students at each grade are provided in Fig. 12.1. It shows four stages of education: primary; lower secondary; upper secondary; and higher education (Ministry of Education, 1992). Elementary schools, middle schools, high schools, and universities constitute the skeleton of the educational system. Civic schools and higher civic schools were established for those who are uneducated and above school age and provide general education, civil education, or vocational education.

The most salient feature of educational development in Korea has been its quantitative expansion. The provision of free compulsory education enabled all relevant age-group children to enrol in school, and the number of secondary-school students tripled during the 1960s (Ministry of

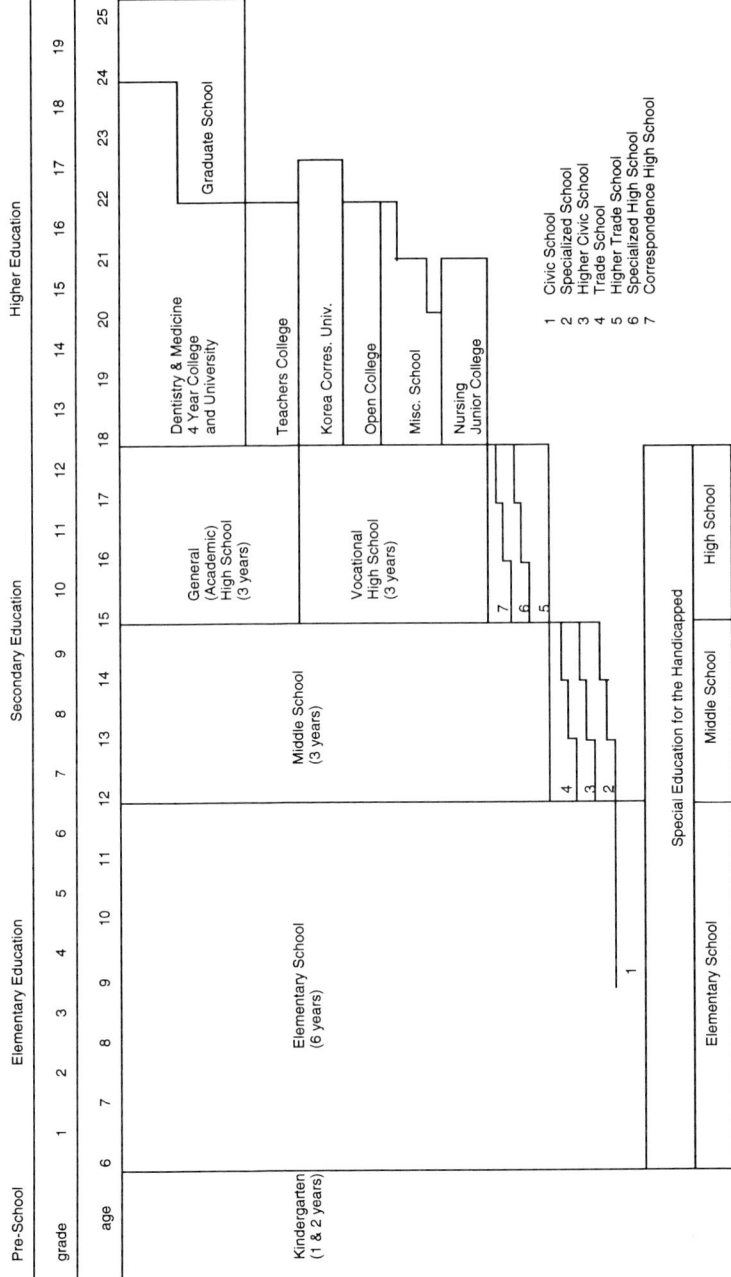

Figure 12.1 The structure of education in Korea: social and educational contexts.

Education, 1992). Entrance to schools on the next level became more competitive.

Education is perceived as a major means of upgrading one's social status. Koreans are eager to educate their children. In 1990, the percentages of parents willing to send their sons and daughters to colleges or higher institutions of learning were 86.3% and 75.7%, respectively (National Statistics Office, 1991).

Each high school is evaluated by the number of students they send to a higher level. Parents want to send their children to prestigious high schools which send more students to universities. The names of high schools which send more students to prestigious universities appear in the newspaper, and such high schools usually get more parental involvement, and financial or moral support from alumni and community. However, there is little official activity, if any, to recognize and reward those high schools for their success.

The critical competition to enter prestigious schools resulted in reforms of lower secondary and upper secondary schools in 1969 and 1974, respectively, intended to make schools more equal (and discussed further below). However, although differences between schools were greatly reduced, all schools are still not equal in their facilities and in the quality of entering students. Differences between schools are even greater at upper secondary level than lower. Students are currently allocated to lower secondary education without any entrance examination, but high schools still select in various ways. Thus, for example, from the 1980s on, special high schools such as science high schools and foreign language high schools have been established for the purpose of educating talented and gifted students in each area. These special high schools have been successful in sending their graduates to prestigious universities, and entrance to them has become very competitive.

Some private high schools in metropolitan areas or large cities where the equalization reform was introduced have succeeded in sending more students to prestigious universities by entrance examination oriented education, and became famous after the equalization reform. The high schools located in districts where the residents' socioeconomic status is high also usually send more graduates to prestigious universities, because of the higher quality of entering students, more private tutoring after school, and entrance examination oriented education. Many people are willing to move into the districts, in spite of the higher cost of residence.

External assessments in Korea

Entrance examinations and assessments of students' academic achievements are the two major types of external assessment used in Korea.

Entrance examinations in particular have been a source of major controversy. National and local assessments of students' academic achievement have been administered continually since 1959. In addition, assessments to obtain data for improving pedagogy and curriculum were implemented during the 1980s.

Assessments for selection

All entrance examinations to lower and upper secondary schools and college and universities fall into the category of assessment for selection. Because of the critical competition to enter higher schools, the entrance examination systems have been an important societal issue.

A brief history of assessments for selection

In the past, schools at lower and upper secondary levels differed considerably in the standards of education they provided, and as a consequence there was intense competition to enter the prestigious schools. Fairness in selecting students by entrance examinations has correspondingly been one of the major concerns of the public.

Entrance examinations to lower secondary schools were administered by groups of schools from 1957 to 1961, by the state from 1962 to 1963, and by district boards of education from 1964 to 1968 (Table 12.1). Schools were also allowed to devise and implement their own entrance examination from 1967. However, a very small number of schools followed the new guidelines and did so. The content and format of the examination was scarcely debated, and experienced little change over the period. Nearly all subjects

Table 12.1 The changes in the entrance examination system to lower secondary school in Korea

Academic year	Entrance examination system
1957–61	No entrance exam, or entrance examination implemented by several schools combined
1962–63	National entrance examination
1964–68	Entrance examination implemented by each district board of education*
1969–	Allocation of students without entrance examination

*Schools were allowed to implement an entrance examination in 1967, and some schools implemented the test individually in 1967.
Source: NIEE (1986b).

taught in elementary school were included in the examination except in the year of 1964 when only two major subjects, Korean and arithmetic, were included. The items of the examination were usually multiple choice format, with four alternatives offered.

The highly competitive examinations for entry into the lower secondary school were abolished in 1969. At this time, prestigious schools at lower secondary level were abolished, and equalization of all schools in terms of teachers, facilities, and funding began to be introduced. All applicants are now accepted and allocated by lottery to schools within their district of residence.

The changes in entrance examinations to upper secondary schools show similar trends to those of lower secondary schools (Table 12.2) but without the final move to abolition. Entrance examinations to upper secondary schools were administered by groups of schools or an individual school from 1958 to 1961, by the state from 1962 to 1963, and by district boards of education from 1964 to 1973 (Table 12.2). Schools were allowed to devise and implement their own entrance examination from 1967. However, a very small number of schools followed the new guidelines and did so at this period.

Since 1974, different implementation procedures have been applied to large cities and rural areas. In metropolitan areas and major cities, where so-called 'equalization reforms' have been introduced, entrance examinations are devised and administered by each district board of education, and given to all students within the area. Students who pass the examination are allocated by lottery to schools within the school district of their residence. In the rural areas and small cities, each school devises its own entrance examination for selection on the basis of competition. This allows

Table 12.2 The changes in the entrance examination system to upper secondary school in Korea

Academic year	Entrance examination system
1958–61	Entrance examination implemented by several schools combined or by each school
1962–63	National entrance examination
1964–73	Entrance examination implemented by each district board of education*
1974–	High-school entrance examination system

*Schools were allowed to implement an entrance examination in 1967, and some schools implemented the test individually in 1967.

Source: NIEE (1986b).

some of them to select superior students and remain prestigious schools. The content and format of the examination have hardly been changed over this whole period. Most lower secondary school subjects are included in the exam, and the format of the examination is multiple choice with four options.

Colleges and universities operate with strict enrolment limits. Because of the gap between college admission limits and the number of aspirants, there are, each year, a large number of students who spend another year preparing to take the entrance examination another time, and this intensifies the competition for entry into tertiary education.

From the mid-1950s to the late 1960s, college entrance examinations were individually devised and implemented by each college, except for two years of state-controlled national examinations from 1962 to 1963 (Table 12.3). From 1969, a college aspirant had to pass two hurdles: a state-controlled Preliminary Examination for College Entrance (PECE) and a main entrance examination administered by each college. PECE was a qualifying examination for college entrance at the outset, and selected one-and-a-half to two times as many students as the college admission limits. From 1973, the PECE score was reflected in the procedures for screening

Table 12.3 The changes in the entrance examination system to colleges and universities in Korea

Academic year	Entrance examination system	Agency
1955–61	No exam or entrance exam by each college	College
1962–63	Qualifying examination for college entrance	CCQECE
1964–68	College entrance examination by each college	College
1969–81	Preliminary Examination for College Entrance (PECE)	CCPECE
	College entrance test by each college	College
1982–93	Scholastic Achievement Test for College Entrance (SATCE)	NIERT NIEE(NBEE)
1994–	Scholastic Ability Test for College (SATC)	NBEE
	College entrance test by each college	College

Note:
CCQECE (Central Committee of Qualifying Examination for College Entrance)
CCPECE (Central Committee of Preliminary Examination for College Entrance)
NIERT (National Institute of Educational Research and Training)
NIEE (National Institute of Educational Evaluation)
NBEE (National Board of Educational Evaluation)
   Source: NIEE (1986b).

students for admission to colleges and universities, and the proportion of the total score for screening determined by the PECE score was gradually increased. The PECE included nearly all the subjects taught at upper secondary level, such as Korean, social studies, mathematics, science, foreign language, industrial art and home economics, and physical ability. The format of the test was paper-and-pencil multiple-choice items with four options, except in the case of physical ability.

In 1982 the PECE was replaced by the centralized college entrance examination, the Scholastic Achievement Test for College Entrance (SATCE). The main entrance examination conducted by each university was abolished in 1981. The SATCE score and achievement in high school were important determinants of eligibility for entrance. From 1986 on, an essay test administered by individual universities was added to the entrance examination system. The subject-matter tested and format of items of SATCE were nearly the same as those for the PECE. Completion and essay items were used along with four option multiple-choice items in SATCE from the 1988 academic year.

After several years of research, a new college entrance examination system was introduced in 1994. The gist of this revised policy was to transform the SATCE into a Scholastic Ability Test for College (SATC), to increase the weight of high-school grades in the process of screening successful candidates, and to extend collegiate autonomy by allowing colleges and universities once again to undertake their own entrance examinations. The SATC was designed to assess higher-order thinking abilities and consists of assessments of verbal ability, mathematics, problem solving, and a foreign language, rather than high-school subjects. All SATC items have five option multiple-choice format. The first SATC was administered by the National Board of Educational Evaluation on August 1993.

Institutions for the entrance examinations

The agencies responsible for the implementation of the entrance examination to lower and upper secondary schools are, for the most part, each district board of education or each school.

The agencies responsible for administering college entrance examinations are more varied. The central and local committee for the qualifying examination for college entrance was established in 1962/3. The central and local committee for PECE had had the responsibility for the implementation of the state-administered college entrance test, PECE, between 1969 and 1981. Each college had individually devised and administered the entrance examination until 1981. The National Institute of Educational Research and Training (NIERT) had the responsibility for implementing

SATCE from 1982 to 1985. The National Institute of Educational Evaluation (NIEE), established in 1985, has administered SATCE since 1986. NIEE was renamed the National Board of Educational Evaluation (NBEE) in 1991, and has been implementing the SATC since 1993.

Major changes of the assessment for selection

Entrance examination systems have been one of the major concerns of parents. Controversies and public dissatisfaction have continued, and successive reforms have been intended by the Ministry of Education to reduce this. Whenever a change in Korea's political structure occurred, the entrance examination system usually became one of the targets of reform.

In 1961, a major change occurred in the political structure, when the military government was established. A provisional measure on entrance examinations to lower and upper secondary schools and college was announced on 12 August 1961. The new measure aimed at the normalization of school education, at fostering the development of provincial schools and universities, and at introducing state-controlled entrance examinations (NIEE, 1986b). As a consequence, state-controlled entrance examinations were implemented at all levels for two years.

However, many problems were identified in the state-controlled entrance examinations. The gaps among schools and universities became greater, school education became more examination oriented, and the weight given to students' scores for physical ability was too high. After two years, the state-controlled entrance examination system gave way to that used prior to the implemented measures.

During the mid-1960s, however, the entrance examinations were again blamed for many educational problems, such as preventing students from normal physical development, examination-oriented school education, excessive private tutoring, and widening gaps among schools. The government announced a new policy for entrance examination to lower secondary school. All schools at lower secondary level were equalized and prestigious schools were abolished. Students who wanted to enrol at lower secondary schools would be allocated to the schools on the school district of residence without an entrance examination.

In the fall of 1968, the government revised the education regulations with regard to the college entrance examination (see Table 12.3). A state-controlled examination, PECE, was proposed. This, as described earlier, was to be a qualifying examination, determining students' suitability to apply for admission examination at universities and colleges. The new policy aimed at solving the problems with college entrance examinations, such as a lower quality of those who were to enter universities, and was also designed to prevent any controversies and irregularities in university entrance procedures.

On 28 February 1973, a new plan to improve entrance examination systems to upper secondary school was announced. As in the 1960s, many critics insisted that too much study to pass the entrance examination (in this case for high school) was interfering with lower secondary students' normal physical development, that school education had become too examination oriented, and that school differences had become greater causing more severe competition to enter prestigious high schools or colleges, and private education expenditure burdens for parents. The new plan aimed at: (i) normalizing lower secondary school education; (ii) equalizing upper secondary schools; (iii) promoting science and technology education; (iv) fostering the development of provincial schools; (v) lessening the educational expenditure burden of parents; and (vi) controlling the concentration of students in the schools in metropolitan areas. This last aim was because many parents were willing to move into metropolitan areas if their children entered prestigious high schools in those areas.

Equalization reform policy was advocated by many people, but others criticized it on the grounds that it would lower the standards of school learning and discourage the education of gifted and talented students. To offset the possible problems of equalization reform policy, special high schools, science high schools and foreign language high schools have been established since the 1980s. The debate about the equalization reform policy still continues.

Another major change occurred in 1980. The Education and Information Subcommittee of the Special Committee for National Security Measures announced a package of drastic measures on 30 July 1980, which was designed to discourage overheated private tutoring and get school education back to normal. The package included the abolition of individually-held college entrance examinations. A 'graduation quota system' which put restraints on the number of college graduates would go into force in 1981, resulting in a hike of college openings. The SATCE was introduced from the 1982 academic year.

Most recently, the Council for Educational Innovation, a presidential advisory committee on education, suggested a new college entrance examination system which was administered from the 1994 academic year. Colleges and universities were to be allowed to select freshmen in a variety of ways, using students' SATC score, high-school achievement, or college and university entrance examinations. The state-controlled examination, SATCE, was replaced by the SATC. However, the university entrance examination administered by many prestigious universities played a major role in screening students. Overheated private tutoring to prepare for university entrance examination and school differences continue to be the subject of public discussion and concern.

Assessment for monitoring in Korea

External assessments whose main purpose is to identify a student-body's level of achievement are categorized as assessments for monitoring. External assessments for monitoring students' academic achievements have been implemented since the late 1950s, nationwide and locally. National assessment data have been collected by several central agencies. Most local centres for educational research have collected assessment data in each district of education since the 1950s.

National and local assessments for monitoring in Korea

Different institutions collected data on national assessment at different times. The National Center for Educational Research (NCER) administered a national assessment every three or four years from 1959 to 1972 (Table 12.4). NCER was abolished and the Korea Educational Development Institute (KEDI) was established in 1972. The Korea Institute for Research on Behavioral Sciences (KIRBS), established in 1968, conducted national assessments for monitoring in the years of 1973 and 1980. KEDI also administered national assessments in 1974 and 1978.

The department of evaluation management was established in the National Institute of Educational Research (NIER) and NIER was renamed the National Institute of Educational Research and Training (NIERT) in 1981. The major task of the department was to devise and implement the college entrance examination, SATCE and national assessment. They conducted foundational research on national assessment in the first year (NIERT, 1981). The department of evaluation management became a separate institution, the NIEE in 1985. NIEE has been in charge of entrance examinations for college and national assessment. A five-year national assessment scheme was planned by NIEE in 1986 (NIEE, 1986c). NIEE (or NBEE) has administered national assessments at primary, lower secondary, and upper secondary level since 1987 (Table 12.4).

Many provincial or municipal centres for educational research have collected assessment data since the late 1950s. In the years of 1972 and 1973, all provincial institutions of educational research cooperated with one another and administered national assessments for 7th through 9th Grades (NIEE, 1986a).

Assessment for monitoring has never become a social issue in the way that assessment for selection has been. However, professional scholars

Table 12.4 National assessment for monitoring students' academic achievement in Korea

Academic year	Agency	Grade assessed
1959	NCER	5th and 6th Grade
1963	NCER	5th and 6th Grade
1966	NCER	5th and 6th Grade
1968	NCER	1st through 6th Grade
1972	NCER	6th Grade,
1973	KIRBS	6th Grade
1974	KEDI	3rd through 9th Grade
1978	KEDI	6th and 9th Grade
1980	KIRBS	6th Grade
1987	NIEE	10th Grade
1988	NIEE	7th and 11th Grade
1989	NIEE	4th through 6th Grade
1990	NIEE	8th Grade
1991	NIEE	4th through 6th Grade
1992	NBEE	11th Grade
1993	NBEE	4th, 9th and 11th Grade

Note:
NCER (National Center for Educational Research).
IRBS (Korea Institute for Research on Behavioral Science).
KEDI (Korea Educational Development Institute).
NIEE (National Institute of Educational Evaluation).
NBEE (National Board of Educational Evaluation).
   Source: KEDI (1975, 1978); NBEE (1991a, 1992); NIEE (1988b).

strongly recommended that national assessment for monitoring be established (Lee, 1982). Seminars on improving and monitoring the quality of primary and secondary education were held during the 1980s (KEDI, 1983; NIEE, 1987b). Many problems concerning national assessment were reported: a lack of long-term planning, a lack of systematic investigation about background and environmental variables, a low level of statistical analysis of the assessment results, and failure to publish technical reports (Jang, 1987). As a result, the use of national assessment data might not be easy.

Some believe that the results were useful in comparing achievement among regions, sexes, and schools. Sometimes the results were used to reprehend the superintendents or principals of low-achieving regions or schools (Jang, 1987). The assessment reports have been provided to policy makers and supervisors in the Ministry of Education or district board of education. However, the way the data have influenced educational policy-making is not clear, and needs investigation.

The use of the results of local assessment seems to differ little from that of national assessment. The major role of local assessment is comparing schools in that area in terms of students' achievements. Schools and teachers have been encouraged to use the assessment result to improve teaching and learning (NIEE, 1987a, b), but it is not clear how this has occurred.

Assessment for improving instruction and learning in Korea

Assessment for learning includes all assessments whose major purpose is improving pedagogy and curricula. This type of assessment is less frequently conducted than other types of assessments. The Korean Educational Development Institute (KEDI) was established for the purpose of educational development and research, and KEDI collected assessment data for learning as well as for monitoring.

A series of small and large scale studies were conducted to assess the effectiveness of the new instructional system of KEDI during the 1970s. The fifth large scale study to assess the effectiveness of the instructional system was conducted in 1979 (KEDI, 1980).

KEDI also conducted national assessments for improving elementary curricula from 1983 to 1985 (KEDI, 1985). The purpose of the assessment was to evaluate elementary curricula, by collecting data on the basic academic ability of the students. The assessment was conducted longitudinally for three consecutive years. The assessment was organized around three basic areas: verbal ability, computational ability, and utilizing material and data (as opposed to subject-matter recall). Inventories to investigate affective domain and home environment were also administered to provide for a more meaningful interpretation of the results.

Conclusion

Entrance examinations to high schools and national assessments are the main external assessments administered in Korea. Assessments for improving instruction or learning have been implemented from time to time. Assessments for selection have dominated other types of assessments.

The intense competition and social controversies regarding entrance examinations have led to frequent changes in the examination systems. They also have been the main obstacles to the use of the results for other purposes, such as monitoring or learning. The test items as well as the

results have been prohibited from being open to public. Several researchers recently tried to find a trend of high-school graduates achievement from
1988 to 1992 based upon the results of SATCE (Park *et al.*, 1994).

Assessments for monitoring have been administered for more than three
decades, and have been the responsibility of four different agencies.
Assessment for monitoring has been criticized for its lack of consistency.
The interpretation and comparison of the results of different assessments
have been difficult. The depth of analysis and the manner in which results
were reported were also insufficient to have an impact on educational
policy-making and preparing materials for supervision.

Compared with other types of assessments, assessments for learning have
not been frequently conducted. Some data were collected by KEDI and
used to improve pedagogy and curriculum development.

School teachers' participation in developing the assessment or analysing the results is very limited. Schools at upper secondary level are evaluated by the number of students they send to colleges or universities.
High-school teachers are expected to teach their students to prepare for
the college entrance examination. Therefore the content and format of the
college entrance examination have a strong influence on high-school
education. The new college entrance examination (SATC) which emphasizes critical thinking abilities was intended to produce innovative reforms
in high-school education.

The research priorities

The new college entrance examination, SATC, is currently the focus of
debate among educators and the public. Many educators, especially school
teachers, have difficulty in understanding the nature of the test. Many
research studies about the pilot trial of SATC, such as item analysis using
item response theory, and differential item functioning (DIF) analyses
(Chu, 1993) have been conducted. Much research is still needed in
clarifying the nature of the test, and in developing appropriate items for
the test, and effective ways of analysing the data.

In the area of assessment for monitoring, a systematic long-term plan
to collect and compare the data should be developed and established.
National criteria for achievement should be developed (Kim, 1991).
Research studies to identify and investigate the background variables
contributing to the students' achievement should be conducted. A case
study of the students' experiences in schools and home may be a starting
point of the research in this area.

Finally, nearly all assessments given to the students so far have adopted
the paper-and-pencil test format (Jang, 1987). Performance tests which are

appropriate to the educational and school contexts of Korea should be studied and developed to assess various educational objectives.

References

Chu, J. (1993). *Identifying Items of Differential Functioning for Sex in Pilot Test for Scholastic Ability Test for College (SATC)*. Unpublished Master's thesis, Ewha Women's University, Seoul, Korea.

Jang, S. W. (1987). The reflections on the national assessment and its use. In NBEE, *The 4th Seminar on Educational Evaluation: The Significance and Direction of National Evaluation of Educational Achievement for the Quality Control of Education*. Seoul: National Institute of Educational Evaluation, pp. 55–75.

Kim, H. (1991). The need and direction of national assessment of educational progress. *Educational Evaluation*, **1**, 69–76.

Korea Educational Development Institute (1975). *1974 National Assessment of Students Achievement Progress: Elementary and Middle Schools*. Seoul: Korea Educational Development Institute.

Korea Educational Development Institute (1978). *Students' Achievement And Characteristics In Elementary School*. Seoul: Korea Educational Development Institute.

Korea Educational Development Institute (1980). *A Research for the Setting the Fifth General Example of the New Instructional System of KEDI*. Seoul: Korea Educational Development Institute.

Korea Educational Development Institute (1983). *The Seminar on Improving tThe Quality of Primary and Secondary Education*. Seoul: Korea Educational Development Institute.

Korea Educational Development Institute (1985). *Evaluation of the Elementary School Curriculum (VI): Comparison of Students' Academic Achievements from 1983 to 1985* (In Korean). Seoul: Korea Educational Development Institute.

Lee, J. S. (1982). The need for national assessment for monitoring. *Newsletter of Korean Association of Educational Research*, **18**(1).

Ministry of Education (1992). *Education in Korea: 1991–1992*. Seoul: Ministry of Education.

National Board of Educational Evaluation (1991a). *A Study on the Evaluation of Scholastic Achievement: Middle School Students*. Seoul: National Board of Educational Evaluation.

National Board of Educational Evaluation (1992). *A Study on the Evaluation of Scholastic Achievement: High School Students*. Seoul: National Board of Educational Evaluation.

National Institute of Educational Evaluation (1986a). *A Study on the Item Development for the Evaluation of Scholastic Achievement: General High School Students*. Seoul: National Institute of Educational Evaluation.

National Institute of Educational Evaluation (1986b). (III) *The history of the change in Korean school entrance examination systems*. Seoul: National Institute of Educational Evaluation.

National Institute of Educational Evaluation (1986c). *The Five Year Development Plan for Educational Evaluation*. Seoul: National Institute of Educational Evaluation.

National Institute of Educational Evaluation (1987a). *Science Achievement and its Correlates: Science Education in Korea*. Seoul: National Institute of Educational Evaluation.

National Institute of Educational Evaluation (1987b). *The 4th Seminar on Educational Evaluation: The Significance and Direction of National Evaluation of Educational Achievement for the Quality Control of Education*. Seoul: National Institute of Educational Evaluation.

National Institute of Educational Evaluation (1988a). *A Basic Study for Development of College Education Aptitude Examination*. Seoul: National Institute of Educational Evaluation.

National Institute of Educational Evaluation (1988b). *A Study on the Evaluation of Scholastic Achievement: Middle School Students*. Seoul: National Institute of Educational Evaluation.

National Institute of Educational Evaluation (1989). *Entrance Examination Policy for Colleges and Universities in Korea*. Seoul: National Institute of Educational Evaluation.

National Institute of Educational Research and Training (1981). *A Basic Research on National Assessment of Students' Academic Achievement*. Seoul: National Institute of Educational Research and Training.

National Statistics Office (1991). *Social indicators in Korea*. Seoul: National Statistics Office.

Park, S., Moon, Y., and Kuh, C. (1994). *Overall evaluation of Scholastic Achievement Test for College Entrance (SATCE)*. A research report under the support of financial assistance of Ministry of Education.

Son, I. S. (1987). *Educational History of Korea*. Seoul: Muneum-sa.

CHAPTER THIRTEEN

Ideals and reality: Sri Lanka's attempts to resolve the roles of educational assessment

Sunderi Kariyewasam

The national system of education in Sri Lanka comprises several cycles—primary, junior secondary, senior secondary, collegiate and higher education (Fig. 13.1). The primary cycle of education runs from Year 1 to Year 5 for students from the age of five years. The curriculum is common for all students and is integrated. The secondary cycle is of six years duration in total, comprising two subcycles of junior secondary and senior secondary. All students follow the same curriculum, leading to the GCE O-level examination. Those who qualify, study for two more years and sit the GCE A-level examination, at the collegiate level. Higher education comprises technical colleges, universities, colleges of education and teacher training colleges.

The non-formal system of education covers adult education given by the non-formal education system under the Ministry of Education and Higher Education, the National Institute of Education (NIE) and other government institutions.

Public examinations are controlled exclusively by the Department of Examinations of Sri Lanka which comes within the purview of the Ministry of Education and Higher Education. Major selection points occur at the end of the senior secondary cycle and the collegiate cycle. At these points students sit the O- and A-level examinations, respectively. There are two language media of instruction and examination—Sinhala and Tamil.

The school network

In 1991 there were 10,300 schools, of varied categories (Fig. 13.2). There are marked disparities among schools. The schools reflect the influence of

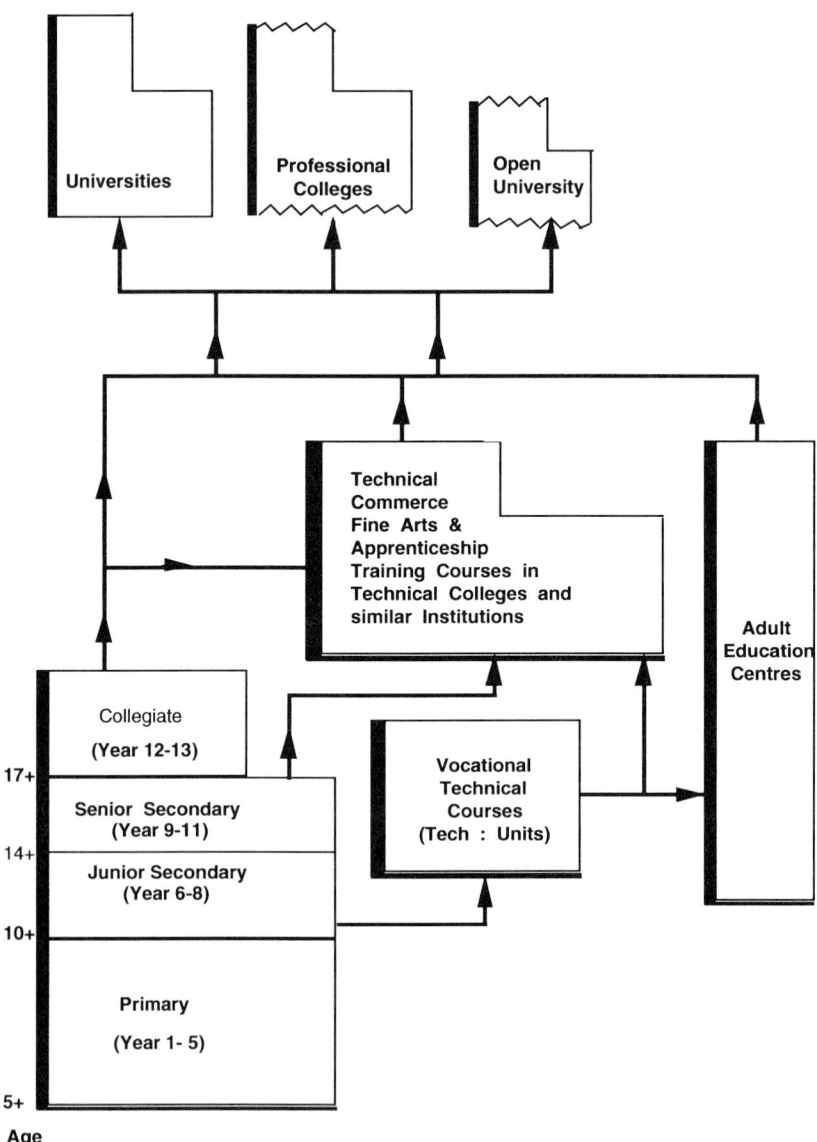

Figure 13.1   The Sri Lankan education system, *c*. 1990.

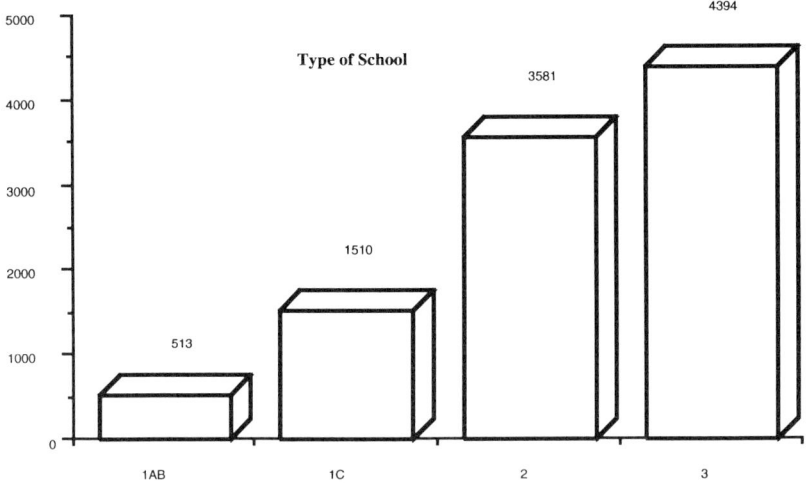

Figure 13.2 Government schools by type, 1991.

geographical, demographic and socioeconomic factors. Every school is classified officially into one of four types:

1 AB: Schools with GCE A-level science classes, arts and commerce
1 C: Schools with GCE A-level arts and commerce classes (no science)
2: Schools with classes up to Year 11
3: Schools with classes up to Year 5 or 8

Sri Lanka has a multi-ethnic, multi-religious population of 17 million. The majority of the people are Sinhalese who account for 74% of the total population, 18% are Tamils (12% Jaffna Tamils; 6% Indian Tamils), 7% are Moors and 1% Burghers.

Agencies responsible for assessment

Assessment is of three main types—student assessment, student selection and large-scale assessments of educational achievement. These interlock in important ways, strengthening each other.

In the first type of assessment teachers identify the educational objective sought. They evaluate the degree to which the desired changes in student behaviour have taken place. By this, final marks are assigned to a student. Educational assessment is now an important activity of the school. The number of teacher constructed tests given annually cannot even be estimated. But they are not scientifically constructed. Hence such results cannot be utilized for evaluation purposes. Internal school examinations are undertaken for the purpose of improving teaching–learning processes, reporting student progress, classifying and placing students within each school and promoting students from one grade to the next.

In the primary and secondary schools, examinations include formative tests at the end of each unit of instruction and end of term summative tests. Formative tests are used for improving teaching-learning processes whereas summative tests are used for reporting students' progress, classification and for promotion from one grade level to the next.

The Department of Examinations is the national body responsible for the second type of assessment, i.e. assessing students for selection purposes and certification. These tests too are not being scientifically constructed and hence the practice of selecting on the basis of such tests can be challenged. It is a common feature in the assessment processes in Sri Lanka.

The third type of assessment monitors educational progress. Before the establishment of the NIE, the Department of Educational Guidance, Research and Evaluation of the Ministry of Education conducted national assessment of educational progress in the 'tool' subjects. The Department of Evaluation of the NIE is now responsible for all aspects of assessment in education. The NIE appraises the effectiveness of the curriculum, instructional materials and procedures and organizational arrangements of the education system. The NIE has also conducted the second stage of the national assessment of educational progress in the 'tool' subjects.

In the early 1980s UNICEF took a keen interest in the monitoring of educational progress. Their studies were attempts to assess and evaluate the achievement of primary-school children at national level from Years 2 to Year 5 in reading and mathematics. These have been conducted with a view to establishing learning situations appropriate to the needs, abilities and potentialities of students of Years 2 to Year 5. UNICEF's aim was to improve through measurement the various elements of the learning situation, such as (a) grouping within the classroom; (b) individual instruction with attention to specific accomplishments and deficiencies; (c) objective assessment of achievements and progress; and (d) a true perspective of the students' errors in understanding and shortcomings. UNICEF also collected assessment data in settlement areas of Sri Lanka. A diagnostic survey of six project areas identified by UNICEF was conducted by using scientifically developed tests. An assessment of student competence in language and mathematics in the six project areas was conducted. The objectives of these surveys were to monitor the students' progress, and provide a feedback to pedagogy.

Use of assessment results to improve pedagogy and learning outcomes

The educational reforms introduced in Sri Lanka between 1972 and 1977 included reforms in assessment. The centralized examination system moved from a selective to a mass basis, which entailed changes in techniques. It also moved to a decentralized mode for some subjects—prevocational studies, physical education and aesthetics education. These were to be assessed continuously and the cumulative marks were to carry weight in selection to higher education.

The main objectives of the 1972 reforms were to bring about social justice and thereby develop the total personality of the child. A number of changes were made. These changes included the use of objective-type tests for evaluation of every subject; the use of continuous assessment in several practically biased subjects; and the use of a five-point scale for analysis of test results.

The White Paper of 1981 made some sound proposals on assessment. They included the monitoring of the gradual development of the student; using syllabuses as criteria of reference; the provision of facilities to develop skills and natural abilities of children; the promotion of individual assessment, classroom evaluation procedures and self-evaluation by the students; and the introduction of a Pupil's Evaluation Record (PER). Evaluation at junior secondary level was to be achieved (a) by continuous in-course assessment or assessment by the school against a check-list; and (b) by an end-of-course examination. Feedback and remedial measures were built into the proposals. Since Sri Lanka was accustomed to evaluation based on public examinations for success and failure, this was a step forward.

In the event the PER system was not implemented successfully for various reasons. These included:

- Regional imbalances and wide disparity in physical resources such as buildings, furniture and equipment.
- Lack of competent teachers to handle a programme of this nature. Their knowledge, skills and practices were so inadequate that it would have taken a quarter of a century to train them.
- Enormous disparities in the socioeconomic status of the students.
- Malpractices in the process of evaluation which favoured certain groups of children.

The White Paper proposed for the junior secondary stage a pupil performance profile which in effect was an extension of PER. At this level continuous assessment was to be carried out on academic subjects; work oriented activities (i. e. vocational subjects, project work, hobbies, etc.); co-curricular activities; participation and performance in sports and games; and creative activities in aesthetic and technical subjects.

This was a lofty ideal. The success of any programme lies in the practicability of implementing it. Continuous assessment in non-academic subjects was a complex problem for educationists, administrators and teachers in a country bereft of staff qualified in psychometrics. The White Paper stated 'guidelines and instructions will be formulated for implementing the system of continuous assessment and teachers will be trained in the procedures and techniques of assessment' (White Paper, 1981, paragraph 18). It further said that assessment in the non-academic areas would be based on graded evaluation schemes developed on commonly agreed criteria with suitable safeguards to ensure the reliability of evaluation methods (ibid, paragraph 20). It did not take into account the practicability of implementing such a scheme. Such a scheme calls for competent teachers of ingenuity and dedication. But the country lacked a sufficient number of teachers to handle the programme.

It also proposed the Grade 8 (Year 9) examination at the school cluster level. The final progress report of the pupil at the end of Grade 8 (Year 9) was expected to contain:

• the grades scored in the final examination (ABCDE)
• the grades obtained in continuous assessment
• performance evaluation with respect to co-curricular activities.

These proposals appear to be very advanced and constructive. Assessment has been widened to include even non-academic areas. But it has not been successfully achieved even in advanced countries. Emphasis laid on continuous assessment is another progressive characteristic of this proposal. This also led to a corresponding reduction in the value of end-of-course examinations. Sri Lankans are obsessed with public examinations and place high value on them. The proposal was never implemented.

A school-based programme of student evaluation was introduced in 1984. The ultimate goal was to translate measurement data into teaching prescriptions. It was based on the complete set of evaluational objectives held by the school. In Sri Lanka testing is confused with evaluation too often. Schools interpret evaluation to mean testing alone. Though an important part of the evaluation programme, testing is still only a part. Parents accept the marks entered in the end-of-term school report. What the marks indicate is meaningless to them. They are either dismayed or satisfied with the marks and that is the end of it. The school-based programme takes into account many other approaches including observation, rating scales, questionnaires and interviews. Evaluation that considers a variety of data can also focus on the process of learning rather than on the product alone. Data collected thus were the basis for evaluating individuals in terms of their abilities and needs.

A number of instruments employing methodologies attuned to identifying barriers to learning, and related to specific instructional techniques were developed. These included programmes with stated objectives arranged in hierarchical order for sequential learning; criterion referenced testing; diagnostic tests to determine strengths and weaknesses; and keyed instructional materials to enable students to focus on deficiencies. These fitted well into the school-based evaluation programme.

Since a programme of evaluation cannot be separated from the total educational programme, school personnel in administration and instruction participated in its inception, organization and promotion. Only the 'tool' subjects of the primary school were included in the programme. It was piloted in one of the regions. They were re-evaluated in order to review pedagogy, reassess intended performance outcomes, re-examine the instructional processes for intended outcomes and to ascertain the type of support services needed to maintain the programme.

The national assessment of educational progress was one of the immediate responses of the Ministry of Education to the changes in the education system in 1985. The ministry was concerned that the educational progress of the country should be monitored systematically and scientifically. The listing of viable learning outcomes serves as an eye opener for the planning of school programmes, selection of curriculum content, and instructional materials and facilities that will serve the needs of the growing child.

The programme was concerned in assessing the tool subjects, namely, mathematics and language at the primary level. The tests were designed to determine what Sri Lankan school children know and can do in language and mathematics at the ages of six, seven, eight, nine and ten years. Underlying this concept of achievement measurement is the recognition of a continuum of knowledge and an acquisition of knowledge ranging from no proficiency at all to perfect performance. A student's achievement level falls at some point in the continuum as indicated by the results of the test. The standard against which a student's performance is compared when measured in this manner is the behaviour which defines each point along the achievement continuum. Along such a continuum of attainment, a student's score on a criterion referenced measure provides explicit information as to what the individual can or cannot do. The measures indicate the content of the behaviour repertory, and the correspondence between what an individual does and the underlying continuum of achievement.

The objectives of the study were:

1. to contribute to a common goal to improve education for children through long-term development activities. This was to enable policy makers, teachers and students to have feedback on the impact of instruction and the progress of students, as a source of information on which to base major curriculum changes
2. to assess the status and changes in educational attainments of children in the primary schools
3. to report long-term trends in educational assessment of children of primary level in Sri Lanka. This recognizes the ongoing long-term nature of the assessment project and the information it will provide. It will be possible to detect long-term trends as well as short-term changes and the existing status
4. to make the data of the national assessment programme available for research on educational issues, policy-making and curricula development,
5. to develop assessment technology through ongoing research and operational studies
6. to disseminate assessment methods and materials, and to assist those who wish to apply them at national, district and school levels.

The results of the study were utilized by policy-makers and curriculum developers and managers of the education system. As a result of the study, curriculum revision, teacher training, the adaptation of teaching methods and materials and the allocation of local educational resources have gone a long way towards coping with pedagogical challenges. The pedagogical challenges involved in introducing formal education to the less privileged are by comparison very much greater. The small schools and the below average schools had been labelled 'deprived schools' and this was closely reflected in the results.

A research programme on the problems of education in disadvantaged primary schools in Sri Lanka has been launched. The majority of the schools in Sri Lanka are small schools and many of them are disadvantaged. Since 1983 the Swedish International Development Authority has placed special emphasis on bringing about a qualitative development in these schools by way of teacher development, improving infrastructure, furniture and equipment. A particular problem facing teachers in these schools is the need to teach several grades of children simultaneously—multigrade teaching.

The prime task of the above programme was to build these schools as viable pedagogic systems. The most important indicator of quality improvement is the level of student achievement, which is yet to be achieved. Studies are being carried out in disadvantaged schools to:

1. develop a series of tests to assess minimum learning competencies expected of children in disadvantaged primary schools in language and mathematics
2. develop research on the organization of syllabi and curriculum material so that they could be reordered to support multigrade teaching
3. design learning material that will facilitate the teaching–learning process among primary children (under the reorganized syllabi)

The above areas of research are of prime importance as they will contribute to improving the teaching–learning process and raising levels of student achievement. Most of these disadvantaged schools have one teacher working in several classes and even the concentrated efforts to appoint more teachers have failed. Multigrade teaching will remain the dominant form of class organization in most disadvantaged primary schools for several years to come. The studies will also enhance understanding among teachers through action research and contribute to a common goal to improve education for children. The implementation of the research studies themselves will bring about positive changes in these schools and their findings will benefit the other schools in the system as well.

Various changes of the last two decades have complicated further the task of teachers and school administrators. Most of these have focused on the management and enrichment of instruction through new techniques or otherwise. Among the reforms with ardent advocates were automatic promotions, non-grading, multi-grade teaching, continuous assessment, pupil evaluation report and pupil's performance report, and national assessment.

Each wave of reform has left some residue of ideas or techniques that continue to exert influence on school practice. Automatic promotions did not affect a qualitative improvement in education. Since the 1970s educationists have struggled to provide frames of reference for reconciling for claims of the learner and the community. Continuous assessment of pre-vocational education confused the teachers because of the enormity of the task, lack of proper teachers and the failure to simulate the condition of the factory in a classroom situation.

In 1981 a go-ahead was given to 'bring order out of chaos' through continuous assessment. But even before it became fully operational it was greeted by a barrage of scepticism and criticism that cast doubt. It ended up as a national disaster, driving the students to the streets as a protest against the concept and practices. Teachers who were unacquainted with the basic concept were unable to do it justice. The whole idea had no scientific base in testing. The subjective element in marking was notorious. Malpractices in assessment crept in. Teachers, pupils and even parents were overtaxed. As a result the students demonstrated against the state. The rebellious movement of the youths made use of this issue to provoke the student population. As a result the state was compelled to abandon continuous assessment.

Some of the scepticism had legitimate roots in deficiencies in planning, management and staffing. But considerable hostile criticism arose, because of the heavy demands on the understanding, diligence and the capacity of the teachers, who were without the knowledge and skills of testing. The result is that today the very term 'continuous assessment' is taboo.

Criticisms of examinations

The harmful role which examinations play in education is emphasized frequently. Examination results are the subject of adverse criticism, even as indices of intellectual values. The country as a whole is ignorant of the folly of deciding selection based on raw marks. No one challenges the validity and reliability of examinations for selection based on raw marks. Highly competitive selection for top-level administration, university entrance and primary Year 5 scholarship awards are decided by tests that are not constructed scientifically. Standardization of marks by language medium at the GCE O- and A-level created considerable tension between Sinhalese and Tamil communities. It was abandoned as a consequence.

The Year 5 scholarship examination

The Year 5 scholarship, which enables successful students to gain admission to good schools, causes a backwash effect on the primary school curriculum and anxiety among pupils, parents and teachers. The NIE is conducting a research study on this examination and a preliminary report suggests the need for a scientific test base for all examinations. The scholarship examination was designed originally to award scholarships to the poor. Since 1985 it has been an open competitive examination due to pressure from many social groups. The function of the examination has changed drastically and now it is the gateway for the affluent to gain admission for their children to leading urban schools. Another criticism levelled against this examination is that the rural schools are affected by it unjustly as their best students are drawn by the more popular urban schools.

Opposition to the examination is mounting and many, including politicians, believe that it needs modification.

Students are required to face uniform educational programmes, but provisions for differences in rate of learning, style of learning, and other characteristics are inadequate. Students are placed in age-graded classes and are expected to attain the same instructional objectives by studying the same graded basic text books. Students are frequently evaluated by teacher-made norm referenced tests of educational achievement, and such tests are often used for improving their instruction. Teachers devote little time to planning and evaluating instructional activities.

Examinations department practices

In the examinations conducted by the Department of Examinations the cut-off marks for a pass, merit and distinction vary. It is the usual practice to pre-decide the percentage of passes etc. and vary the cut -off points accordingly. It is the usual practice to adopt conference marking. Studies have revealed that there is considerable variation in the marking of scripts although every effort is made to guide the examiner with detailed marking schemes.

University entrance

Admission to the universities is determined by success at the GCE A-level examination conducted by the Department of Examinations. There is great concern regarding the number admitted every year. Over the years the ratio between the number admitted and the number who seek admission has worsened. Those who succeed in gaining admission face severe competition and great hardships. The most depressing feature has been that the actual numbers admitted have increased only marginally over the past few years, despite the fact that a large number meet the entry requirements and are officially 'qualified' to enter.

These restricted admission policies have created an explosive situation among students, parents and administrators. It has caused ethnic tension in the country. As a result standardization of marks between different language media was proposed. This in turn caused an uproar and was abandoned. Limitations of places in the universities have resulted in frustration for thousands of qualified youths with no other avenues for future education. The district quota system was introduced to combat the impoverished environmental influences on the poor and underprivileged. This is not based on any rational criteria. Currently the University Grant Commission (UGC) continues to select students on the basis of: 'all island' merit 30%; on district-basis merit 55%; and the balance of 15% on underprivileged district merit.

Admissions are based on the aggregate of raw marks. The problem of admission to university has become a political hot potato for Sri Lanka. A special GCE A-level examination was held in the north and the East in April, 1991, in place of the August 1990 examination that could not be held because of violence prevailing in the north and east in Sri Lanka. But the

UGC later decided to consider the two examinations separately for university admission. Students in the north who aspired to enter the engineering and commerce faculties protested. They planned legal action in the Supreme Court against the UGC, alleging discrimination in university admissions. The court order was that the results of the two examinations should be considered as one for admission purposes.

The emergence of alternative assessment procedures

These developments resulted in the creation of the department of evaluation in the NIE with technical and pedagogical functions. It attempted mainly to resolve some of the tension that spread in the country. Within the ministry bureaucracy there was almost a constant conflict among opposing view points. This department undertook the specific task of in-service training to educate the teacher population in assessment techniques to improve learning. They also organized in-service training for system monitoring.

Officers of the department of evaluation of the NIE continued to use the results of the national assessment of educational progress together with statistical data on attendance, repetition, promotion and drop-out to improve the curricula, instructional methodologies and practices. They submitted an organizational structure proposal for the execution of their work. In every stage they have preserved the matrix form of each organization structure, particularly in its technical and logistical functions.

Logistical and administrative areas

All the administrative procedures and the achievement of its objectives are centred in this unit, defining and coordinating the financial and material resources required to fulfil the activities and processes of the technical unit. The major duties of the unit have included the general planning of activities for all the stages and work groups; the appointment of professionals for specific tasks, teachers for item writing and evaluation committees for these items; the determination and control of security standards for the tests for all the stages; and the design and supervision of packaging, transportation and reception of materials.

Educational area

The use of assessment data for pedagogic purposes has been gradually increasing among teachers and schools. The commitment by the Ministry of Education and Higher Education to the use of the data for the formulation of education policies, has been very encouraging. The ministry has promoted a pedagogic programme (curriculum renewal) and training for teachers in schools of deprived areas, whose results are extremely low. The results have also served as a basis for the review of study plans and programmes and curriculum development for the whole country.

Technical area

This area is responsible for the execution of all the technical procedures required for the assessment instruments, and the analysis of the results of the assessment carried out year after year. Accordingly, this area is in charge of a subject group. They are in charge of the development and improvement of the assessment procedures of the achievement of educational objectives. In the preparation of assessment instruments the jobs in this area are subdivided according to learning fields and the content of the tests.

The main responsibilities of the technical area have been the following:

- Selection of learning objectives.
- Selection and training of item constructors, preparation of guide lines for the elaboration of items and criteria for their evaluation.
- Setting up of evaluation committees for item construction, organization and conducting of work sessions.
- Assembling of trial versions of various tests.
- Sample designing for pre-testing.
- Statistical analysis of the test items.
- Assembling of final tests and estimates of reliability.
- Preparation of support and dissemination materials, technical brochure for teachers and manual for pedagogical orientation.
- Designing a national dissemination plan.

The teacher's role in assessment

In view of student heterogeneity, the teachers' role in expediting the final educational process becomes increasingly difficult. The teachers' role in assessment for learning had been to identify the educational objectives sought. They should know the nature of the student behaviour to be displayed when the educational objectives have been achieved. They must determine what educational experiences the student must have in order to achieve the objectives. They must know their students so well that they can design and order the educational experiences to incorporate varied interests, aptitudes and prior experiences. Finally, they must evaluate the degree to which the desired changes in student behaviour have taken place.

The Fifth Standard Scholarship examination has a backwash effect on the curriculum. From the very inception of its school career a child is so oriented to this examination that teachers neglect the teaching of other subjects in preference for the teaching of mathematics and language. It has affected their new general role.

Because of its competitive nature, the GCE O-level examination encourages students to resort to private tuition. It has been carried out to such an extent, that the role of the teacher is also altering. Those teachers who give private tuition tend to diminish their role as guide and educator. Those who do not give private tuition fail to fulfil their duties, by expecting the students themselves to learn by tuition. Thus, although assessment is capable of being used in Sri Lanka for pedagogic purposes, its role in selection continues to dominate the educational process.

CHAPTER FOURTEEN

Grappling with heterogeneity: assessment in Indonesia

Jahja Umar

Introduction

Indonesia is an archipelago of some 17,000 islands, covering three time zones between Asia and Australia and between the Indian and Pacific Oceans. The largest of these islands are Sumatra, Java, Kalimantan, Sulawesi and Irian. There are 27 provinces, each subdivided into *Kabupaten* (districts) and *Kota Madya* (cities).

The population in 1990 was around 180 million people, distributed very unevenly. Java and Bali have the highest population density, with the former accounting for 60% of the total, and Irian Jaya and Kalimantan the lowest. There are hundreds of different languages and dialects, with Bahasa Indonesia as the national language playing an important role as the language of unity. Indonesian economic growth has been, and continues to be, rapid.

The Indonesian education system

A three-level education system for Indonesia was stipulated by Act No. 2 of 1989: Basic Education, Middle Education and Higher Education. Basic education comprises kindergarten (preschool, referred to as TK), primary school (SD) and junior secondary school (SMP). At the middle level are senior secondary schools, consisting of general senior secondary schools (SMA) and vocational/technical secondary schools (SMK). Higher education comprises universities, institutes (of technology, art, agriculture, teacher training), Academies and *Sekolah Tinggi* (such as schools of economics). This structure is summarized in Fig. 14.1.

Kindergarten is not a pre-requisite for entry to SD. Most kindergartens are private institutions but there are government guidelines for programme

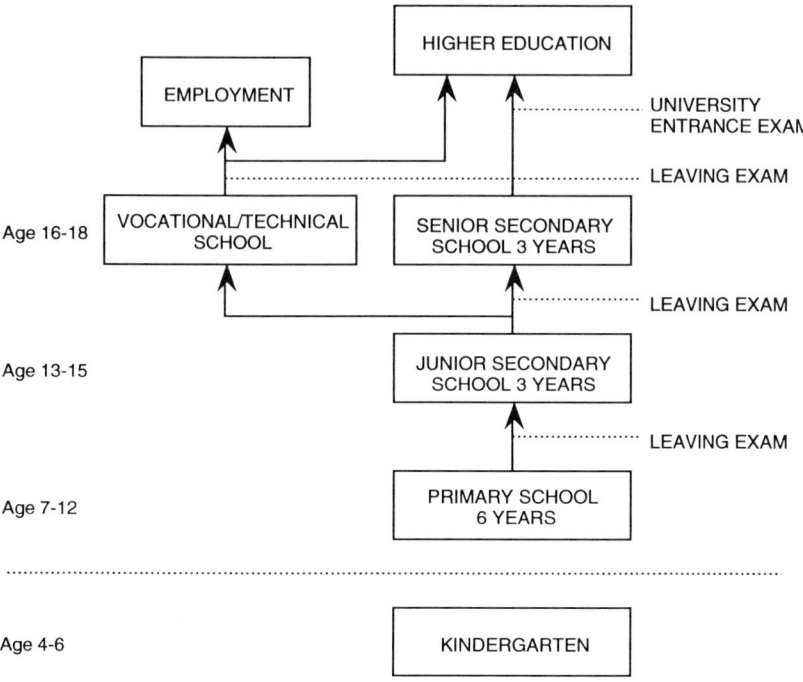

Figure 14.1 Structure of the education and selection system, Indonesia.

content and methods which emphasize socialization and character build-ing, especially in moral education, ethics and religion. Most children in kindergarten are four or five years old.

SD consists of six grades, with entry to Grade 1 being allowed from the age of six. There is no selection to this level of schooling. The school year runs from the middle of July to the end of May, in a quarter system. An effective universal primary education campaign during the 1970s led to a massive increase in primary-school enrolments. Nowadays virtually every young child enters primary school, but drop-out remains a problem, particularly in the more isolated parts of the Republic. Currently there are about 150,000 primary schools with 30 million students and 1.2 million classroom teachers. There is a common and relatively prescriptive curri-culum for all primary schools, but a new national curriculum, which is much less prescriptive and has a 20% local component, will be implemented progressively from July 1994. The majority of primary schools are govern-ment-run, although about 10% are Islamic Schools under the Ministry of Religious Affairs. There is a school leaving examination at the end of Grade 6 (which will be abandoned soon), leading to the award of a certificate.[1]

About 66% of SD graduates continue SMP, which spans three years. In principle, there is no selection for entry to SMP but some popular schools have to select their new students, using test scores from the SD leaving examination. Currently, there are slightly more private SMPs than public. Since the government started a nine-year universal education programme, the enrolment of thirteen- to fifteen-year-old children in SMP is expected to increase dramatically over the next ten years. The national curriculum for the SMP is to be integrated with the SD curriculum in 1994, as a single package for a nine-year basic education. Unlike those in SD, teachers in SMP are subject rather than general classroom teachers. In the past there were vocational/technical schools at junior secondary level. These have been abandoned, but an effort is made to provide students who might not continue their studies with useful practical skills. At the end of SMP Grade 3, students take an examination which leads to the award of a certificate.

The senior secondary level covers three years and consists of two streams: the SMA and the SMK. With its emphasis on an academic programme, the SMA is conceptualized as a preparation for higher education, while the SMK emphasizes skills training for direct entry to the labour market. Act No. 2 of 1989, however, allows SMK graduates to go on to higher education when suitably qualified. There are about twice as many private SMAs as public. A national examination is taken at the end of SMA. Currently, there is no policy on selection to SMA and SMK but, as with SMP, some popular SMA select students using their scores in the SMP final examination. Since SMA is intended as a screening for candidates for higher education, a selection policy and programme are to be developed. In the second and third year of the SMA course, streaming takes place into one of four programmes with the following emphases:

A-1 mathematics and physics
A-2 chemistry and biology
A-3 social studies and economics
A-4 languages and literature.

A new curriculum will be launched in 1994, in which streaming will take place only in the third year.

Higher education institutions, of which there are about 20 times as many private as public ones, offer degree and diploma programmes. The diploma programme takes from one to three years. The degree programme has three levels:

Stratum-1 (undergraduate), usually taking four years
Stratum-2 (master's), taking two years
Stratum-3 (doctoral) taking from three to five years.

There is a selection procedure for entry to the diploma programme and to each level of the degree programme. A senior secondary school graduate wanting to enter a public university or other institution faces the most severe competition: every year, about 600,000 candidates compete for about 70,000 places in public higher education institutions. Private universities operate their own selection processes, and a few of them are also quite competitive.

The Ministry of Education and Culture consists of four operational and three supporting units. The operational units are the directorates general of

- primary and secondary education
- higher education
- out of school education, youth and sport, and culture.

    The supporting units are the

- secretariat general,
- inspectorate general, and
- office of education and cultural research and development.

The assessment system

In principle, there are three types of assessment in the Indonesian pre-university education system:

- assessment by the teachers in the classroom
- external assessment either by a provincial office of the ministry (KAN-WIL) or the director general for primary and secondary education (DGPS)
- national surveys of student achievement, for curriculum monitoring and evaluation.

    In addition there are entrance examinations for students wanting to enter higher education institutions.

Assessment in the classroom

As part of normal teaching–learning activities, a teacher must carry out continuous evaluation in the classroom. Such evaluation might take place after every teaching unit has been completed as well as at the end of each quarter/semester or academic year. The academic year is organized in a quarter system at the primary level, and a semester at junior and senior secondary level. Beginning in the 1993/4 academic year, however, a quarter system will be implemented at all levels.

The purposes of classroom assessment include:

- Monitoring students' progress against the stated instructional objectives for understanding of topics taught.
- Self-evaluation of teaching effectiveness in specific topics so that teaching can be improved or the topic repeated when a large proportion of students fail.
- Identification of learning difficulties in individual students.
- Periodic collection of data on student achievement for continuous evaluation of performance. By combining these data with results from final examinations at the end of a quarter/semester, a report can be made on student performance in that quarter/semester. At the end of an academic year, reports from all the quarters/semesters during that year are used as the basis for decisions on student promotion to a higher grade.

Directions to teachers on how to carry out classroom assessment are provided in a supplement to the curriculum booklet, entitled *Major Guidelines of the Teaching Programme*. These guidelines are provided for teachers at all school levels. The classroom assessment includes all homework, assignments, performance tests, and day-to-day examinations. With this type of assessment, there should be no mismatch or tension between teaching and testing, and there is obvious potential for feedback to teaching and learning.

External assessment

There is a school leaving examination at the end of each school level. The main purpose of this examination is to assess each student's overall achievement so that a decision on certification can be made. In Indonesia, certification for all levels of schooling is based both on performance in the school leaving examination and reports in the last grade. For each student, the average achievement level recorded in the quarter/semester reports for the latest year is combined with his or her performance in the school leaving examination. The result is expressed as a grade on a scale from zero to ten. A student who scores 5.5 or better after averaging across subjects is considered to have passed and is awarded a certificate. This rule applies to all schools in the primary and secondary levels. With the introduction of a nine-year basic education programme in 1994, however, the government plans to abolish the school leaving examination at the end of Grade 6.

Scores from school leaving examinations are further used for selection purposes. In many places parents prefer to send their children to public rather than private schools, because of the lower cost. Since there are fewer places available than the number of applicants, a selection procedure is necessary. Since some of the better schools are more popular than others, it is important that the selection process for these schools is seen to be fair. Furthermore, some schools wish to maintain their exclusive public image and must, therefore, be seen to be selecting only the best students. In practice, secondary schools select their students only on the basis of scores in the school leaving examination at the lower level. Junior secondary schools will select using scores in the primary leaving examination and senior secondary schools will use scores from the junior secondary leaving examination. Reports from classroom assessment are not considered to be a fair instrument for use in these selection processes.

Yet another use for data from the school leaving examinations is for monitoring and comparison of performance schools, districts and provinces. Officials at district and provincial levels of the ministry may issue annual reports on performance at school, subdistrict or district level, sometimes ranking them, and make use of the data as an input for planning and decision making on resource allocation.

Since results from school leaving examinations are used for selection and quality monitoring, high levels of objectivity and comparability in the scores are necessary. For this reason, control of these examinations is made external to the schools. The process of preparation of the school leaving examinations differs with school level and accounts of this process at each level follow.

Primary leaving examination (end of Grade 6). The DGPS establishes a national committee which in turn appoints a national team to construct a test specification table, based on the national curriculum for primary education. This test specification table is sent to provincial offices of the ministry (KANWIL). Each KANWIL then assigns a group of senior teachers who have been trained in item writing to construct items according to the specification table. The training of item writers is provided by the national Examination Development Centre. Usually, each KANWIL develops up to seven forms of the test for each subject being tested. Up to five of the seven versions are administered to students, with schools in different subdistricts using different ones. There is no piloting of items prior to use and no equating of scores across test forms. Since all the test forms were developed using the same test specification, it is assumed they are equivalent. In the future, it is planned to develop a precalibrated item bank so that only high quality, pretested items will be used, and empirical equating or scale conversions between test forms will easily be possible.

In provinces with a very large school system—such as East Java, with a population of 34 million—item writing teams are established at district rather than provincial level. Except on language papers, all items on the school leaving examination are multiple choice, so the administration of the tests in such large provinces is usually by machine-readable answer sheet. In most provinces, however, scoring is done by groups of teachers at the level of small clusters of schools.

Junior secondary leaving examination (end of Grade 9). The procedure for preparing examination materials at this level is similar to that described for the primary leaving examination, except that item writing should take place at provincial level only. Hence, it could be considered as a provincial examination using a national table of test specifications based on the national curriculum for junior secondary education. The examination at this level is closer to being based on an item bank than is the primary leaving examination. For example, in 1993 five provinces included common items in their test forms in some subjects. These common items were taken from a calibrated item bank that is being developed at the national examinations centre (Choppin, 1981; Umar, 1990, 1993). Scores from different test forms can be converted to a national scaled score using item response theory (Hambleton and Swaminathan, 1985; Lord, 1980). In the future it is planned to develop the junior secondary leaving examination using a national item bank network organized by the examinations centre. The development plan for this network is presented in the final part of this chapter.

Senior secondary leaving examination (end of Grade 12). There is greater central control over the preparation of this examination than over those at other levels. First a national committee appoints a national team to construct a test specification table based on the national curriculum. Each KANWIL in the provinces then appoints a team of well-trained item writers to construct items based on this specification table. Items from the provincial teams are sent to a national team of item reviewers who select the best items to construct up to seven approximately equivalent test forms. If there are less than seven good items for a particular item number in the specifications, the reviewing team will modify items proposed by the provinces, or construct new ones, to ensure seven test forms can be constructed. Five of these seven test forms are sent to all provinces and each KANWIL may choose one, two or three out of the five as its senior secondary leaving examination. Once again, the items are used without prior piloting, and there is no equating or converting across test forms. The Examinations Development Centre is now proposing a plan to decentralize the preparation of this examination, using an item bank instead of the present *ad hoc* system of item development. Unlike those from the leaving examinations at primary and junior secondary levels, the results from this examination are not used for selection to the next highest level.

Survey of student achievement

For the purposes of monitoring and evaluation at the national level, there is a need for national data on student achievement at each level of education. Results from school leaving examinations provide information only for Grades 6, 9 and 12, while reports from classroom assessments cannot be compared since they do not use a common scale. In order to obtain such national data, periodic surveys (on a sample basis) of student achievement and other relevant variables must be carried out. Data from such surveys could be used to judge the health of the system, and as a basis for decision making on resource allocation, curriculum development and other planning concerns. If the surveys are organized correctly, longitudinal comparisons of educational data, such as achievement, could also be made. In the existing system, periodic surveys of student achievement at primary and secondary levels are carried out by the DGPS, but these are not well designed and their aim is limited to obtaining information on those curriculum topics in which most students perform poorly. There is no analysis to associate this poor performance with input or process variables in the education system, or with previous educational policy decisions. The quality of instruments developed for the survey, including the achievement test items, is generally poor. A plan for the development of an Indonesian National Assessment Programme (INAP) is now being produced at the Education Development Centre, and an account of this plan is presented later in the chapter.

University entrance examinations

Reference has already been made to the intense competition for places in public, rather than private, universities and other institutes of higher education, and to the use of entrance examinations for the selection process. These examinations are under the complete control of higher education personnel. Indonesian public universities are grouped into three regional clusters: west, central and east. Each region develops its own examinations and the dates of these are arranged so that a candidate may apply to only two courses or universities in the same region. The examinations have negative backwash effects at the pre-university level that will be considered later in this chapter. The level of difficulty of these entrance examinations is very high and, coupled with a great range in the degree of competitiveness among courses and universities, this renders their validity rather questionable. As the items are so difficult, test scores do not discriminate well among lower ability candidates. Score differences may reflect guessing of multiple choice item answers rather than real differences in ability, and the resulting ranking is not valid for selection purposes.

Problems in the existing system

As outlined above, the Indonesian assessment system comprises four distinct activities: classroom assessment, external assessment, national surveys of achievement and university entrance examinations. Although open to criticism on the grounds of economic inefficiency, this multiplicity of assessment forms should increase the validity of the information obtained. As Mehrens and Lehmann (1984) have pointed out, no single assessment activity can satisfy all educational purposes at once and no achievement test can measure everything. Each of the forms of assessment described above has its own strengths and weaknesses. What follows is an account of some of the problems experienced with each form and the efforts that are being made to deal with them.

Problems with classroom assessment

The advantage of using teacher-made tests is the match between teaching and testing. The instructional validity of such tests should be high (Popham, 1978). There are, however, three major problems with classroom assessment:

- the quality of items used is generally poor
- scores from different classrooms or schools are not comparable
- there is a tendency among teachers to over-use multiple choice items that are readily available from books.

The problem of poor item quality can be solved if a large collection of good items is made easily accessible to teachers. There is a need for an item bank containing various types of item which match the curriculum objectives. The development of a good item bank raises problems, however. First, there is a need for well-trained item writers and reviewers who will construct items continuously as a routine. Secondly, a large item bank requires computers and specifically developed software for item banking operations. Thirdly, there should be expertise available for items to be calibrated so that their use may be extended beyond everyday classroom assessment. Once a good item bank is established, however, teachers can share the items and item quality in classroom assessment will, in general, be better. Calibrated or not, a good item is still a good item.

In order to make high quality items available to teachers, the Examinations Development Centre of the Ministry of Education and Culture is now developing a network of provincial item banks. Activities began with the intensive training of item writers and reviewers in the provinces. More than 3000 senior teachers have now been trained and the training continues with teachers from primary, junior secondary and senior secondary schools. After training, these teachers are assigned to write ten items a month which are sent to a KANWIL item bank coordinator for a team to review and revise when necessary. Good items are then sent to the Examinations Development Centre item bank team for field testing and validating before being stored in the bank under two categories: those which fit the Rasch model (Wright and Stone, 1979) are stored in the calibrated item bank, and those which do not fit the Rasch model are stored in the non-calibrated item bank. Both calibrated and non-calibrated items may be used by teachers since calibration is not a concern in classroom assessment. A copy of the Examinations Development Centre item bank file is sent to each KANWIL so that it is accessible to teachers. These activities in the item bank development are summarized in Fig. 14.2.

When comparability of test scores from different schools is required, items from the calibrated item bank can be used so that a score conversion table can be constructed. To deal with the problem of overuse of multiple-choice items, teachers are instructed to use only free-response items in classroom assessment.

Problems with external assessment

Given the wide economic and cultural heterogeneity of Indonesia, it is unlikely that any single, centralized test will be simultaneously valid for all schools. Some items in such a test might function differently in different places. This is especially true with regard to the level of difficulty of the items. If a test is to be taken by students from schools of highly variable quality and resources, then items covering a very wide range of difficulty should be included. This means that some items would function well only in poor schools and others only in good schools. Furthermore, test scores are not comparable unless weighting of items according to their difficulty takes place. Two approaches are being adopted to counter these problems: the decentralization of school leaving examinations, and the exclusive use of precalibrated items in such examinations, to allow the comparison of scores across schools, districts or provinces.

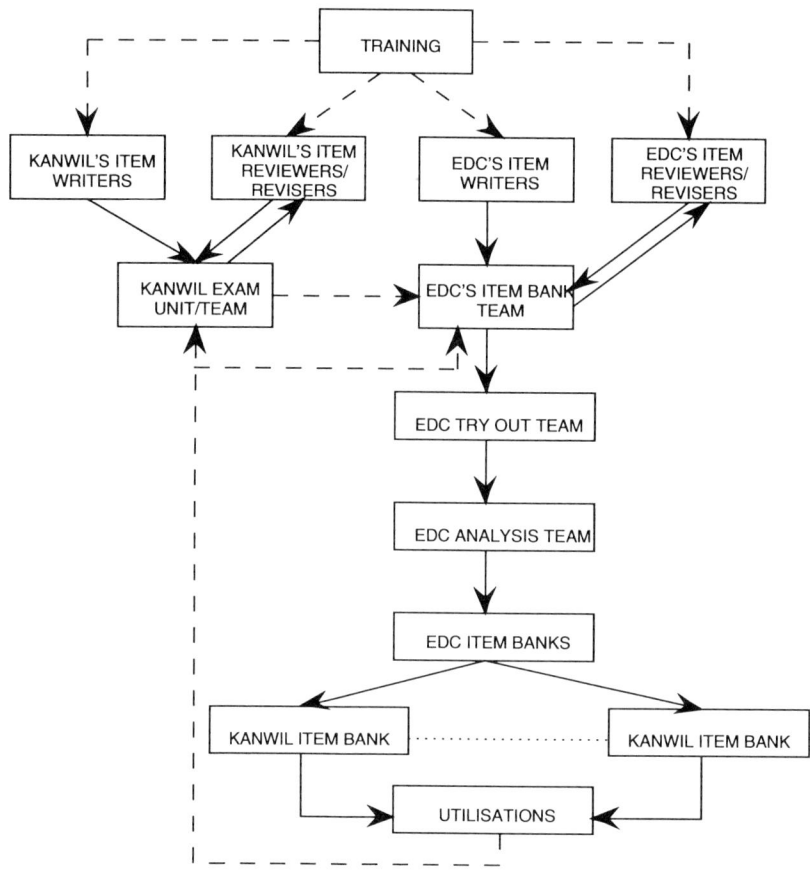

Figure 14.2 Development of calibrated item bank networks.

The present management of school leaving examinations by an *ad hoc* committee presents a problem of possible discontinuity in the system. Development is also made difficult since an evaluation of the system would require another *ad hoc* committee. Figure 14.3 presents a plan for an evolutionary move to an institutionalized system, as an attempt to overcome these problems. The figure shows that future school leaving examinations will be decentralized to the KANWILS or districts, and that a calibrated item bank network will be the source of items used in the examinations.

Problems with national surveys of achievement

The annual national survey carried out through the DGPS for monitoring

1a. Existing Ebtanas Cycle (ad hoc)

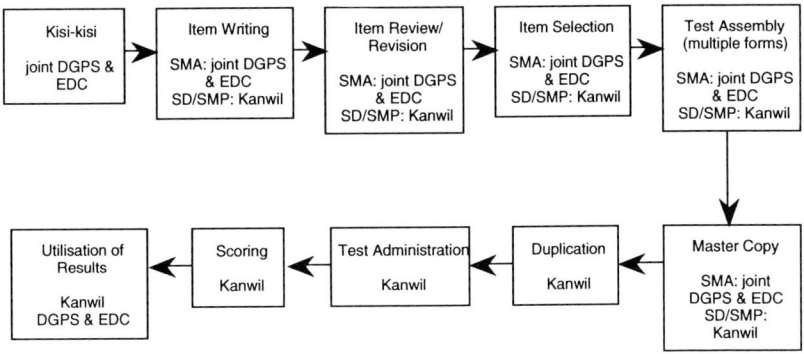

2a. Item Bank Preparation (Institutionalised and coordinated by EDC)

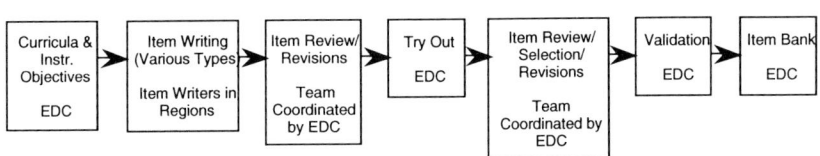

2b. Future Final Exam Cycle: (Institutionalised, coordinated by permanent committee chaired by DGPS)

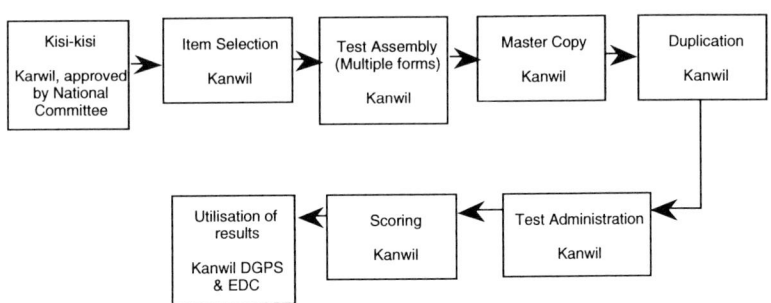

Figure 14.3   An evolutionary shift from the existing system (with *ad hoc* exam material preparation) to item bank-based examination (institutionalized exam material preparation).

purposes can be criticized for its use of poorer quality items than those in the leaving examinations. These items are not constructed by trained item writers, nor do they undergo empirical validation before use. Furthermore, there is no collection of data on educational input and process variables that would permit analysis of possible causes of failure or inefficiency. Longitudinal analysis and improvement or evaluation of the system are made difficult by the *ad hoc* basis of the surveys. It is proposed that an institutionalized systematic monitoring system be established at national level, together with an information management system to assist with policy analysis and decision making. Figure 14.4 presents a design for such a monitoring system, to be known as the Indonesian National Assessment programme (INAP).

Problems with university entrance examinations

A selection system for higher education should meet four criteria:

- fairness
- predictive validity
- economic efficiency
- no negative backwash on preuniversity education.

There is a serious negative backwash effect from the present university entrance examinations. Many teachers at senior secondary level are preparing their students for these examinations rather than implementing the curriculum. Students themselves invest much time and money in test preparation, supporting a rapidly growing private tuition sector aimed specifically at these entrance examinations. There is little that can be done to improve this situation since university entrance policy is completely in the hands of the universities themselves, untouchable by either the Examinations Development Centre or DGPS at the present time. The Examinations Development Centre is, however, trying out a selection system based on 'academic potential' or 'scholastic aptitude' rather than on testing students' academic knowledge as the present examinations do. This approach is expected to be fairer to talented candidates from less developed schools.

Future directions

As this chapter has indicated, the tendency is for the Examination Development Centre to play a more active role in managing the various types of assessment. This centre supports research and development activities in the fields of model development, instrument development and methodological adjustment. If the plans outlined in this chapter are implemented, the Examinations Development Centre will retain a crucial role in assessment in Indonesia.

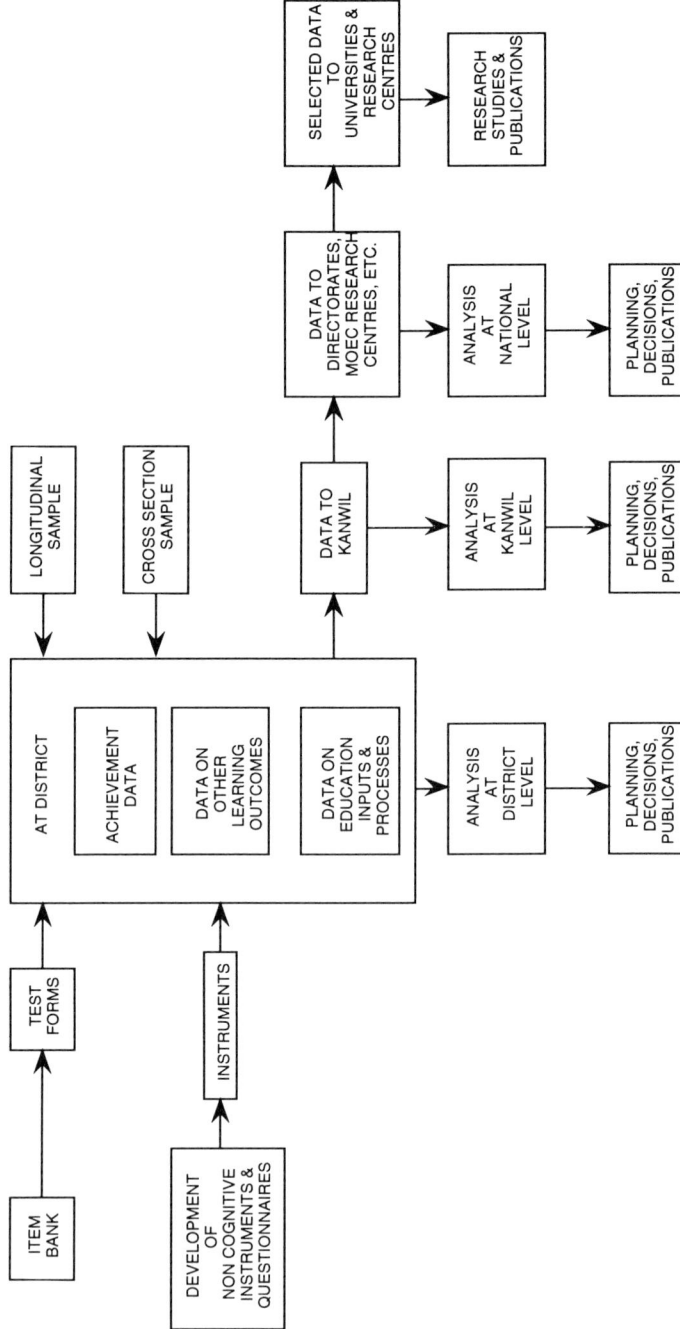

Figure 14.4   Design for future Indonesian National Assessment Programme.

Note

1. This chapter was drafted before the introduction of a nine-year basic education programme to replace the six-year programme. Although this change has now been implemented, it is not yet clear what effect it has had on the primary leaving certificate which was taken at the end of Grade 6. It seems likely that selection pressures will remain between Grades 6 and 7 and that, as a consequence, some form of selection examination will remain in use, for entry to the more popular schools at least.

References

Choppin, B.H. (1981). Item Banking Development. In De Gruijter, D. N. M. and van der Kemp (eds) *Advances in Psychometrics and Educational Measurement.* London: John Wiley.

Hambleton, R. K. and Swaminathan, H. (1985). *Item Response Theory: Principles and applications.* Boston: Kluwer-Nijhoff Publishing.

Lord, F. M. (1980). *Application of Item Response Theory to Practical Testing Problems.* Hillsdale, NJ: Lawrence Erlbaum Associates.

Mehrens, W. A. and Lehmann, I. J. (1984). *Measurement and Evolution in Education and Psychology* (3rd edition). New York: Holt, Rinehart and Winston.

Popham, W. J. (1978). *Criterion-referenced Measurement.* Englewood Cliffs, NJ: Prentice-Hall.

Umar, J. (1990). *Development of an Examination System Based on Calibrated Item Bank Networks.* Unpublished Project Report, SIDEC, Stanford University.

Umar, J. (1993). Item Banking. In Husen, T. and Postlethwaite, T.N. (eds) *International Encyclopaedia of Education* (2nd edition). London: Pergamon Press.

Wright, B. D. and Stone, M. (1979). *Best Test Design.* Chicago: MESA Press.

Part four
RESOLVING THE TENSIONS? POSSIBILITIES AND LIMITS

The final chapters of this collection look to the future. They offer concepts and frameworks for thinking about assessment in a productive way; one which may help us avoid some of the problems of the past, and reconcile more successfully the different objectives of assessment. Caroline Gipps, in particular, is optimistic in her discussion of how far this is possible. Drawing on wide experience of assessment in, especially, western primary schools, and on modern psychological theory, she argues the case for effective 'assessment for learning'. Anthony Somerset's wide-ranging experience of developing countries has made him very aware of the negative effects of examinations in a highly competitive labour market; but also of their potential power for good. If we are clear about the way that education systems affect behaviour then, he argues, examinations can exert a major positive influence on the quality of education. Alison Wolf's focus is on the insights offered by economics and game theory and how they help explain individual behaviour, and some of the major disappointments of previous reform attempts. Nonetheless, here too a fuller and clearer understanding of how incentives operate provides guidance for the future. Reforms which are realistic about the incentives to which people respond, and which harness rather than ignoring them, can be much more effective than in the past.

CHAPTER FIFTEEN

Assessment for learning

Caroline Gipps

Introduction

This chapter is about assessment for learning; my thesis is that assessment for selection, monitoring and accountability purposes—indeed any assessment using a 100% sample—can, and must, be based on a model of assessment which will enhance and support good learning. Unless we do this we will never raise true educational standards. I shall deal with five issues: fitness for purpose, the legacy of psychometrics, learning theory, educational assessment and an agenda for progress. I believe that we know much more than we think about how to design assessment to support learning.

Fitness for purpose

Assessments (which I use here to include tests, examinations, practicals, coursework, observations and records of achievement) come not only in a range of forms but with different purposes and underlying philosophies; these relate to the issue of fitness-for-purpose. For example, assessment to support learning, offering detailed feedback to the teacher and pupil, is necessarily different from assessment for monitoring or accountability purposes. We must first ask the question 'assessment for what?' and then design the assessment programme to fit.

I take the view that the prime purpose of assessment is professional: that is assessment to aid the teaching/learning process. But, government, taxpayers and parents also want to know how the education system and individual schools are performing and must have access to such information. A major element of this information is pupil performance as measured by tests and examinations. Assessment carried out for these purposes is likely to be more superficial and needs to be more 'objective' or reliable than that to support learning.

The problem that assessment has to face is that tests designed for purposes other than to support learning (e.g. the huge quantities of multiple choice standardized tests in the United States, and the formal written exam in the UK), have had unwanted and negative effects on teaching and the curriculum. The stultifying effect of public exams on the secondary system in England was pointed out by Her Majesty's Inspectors (HMI) (HMI, 1979, 1988) and was a prime mover in the shift towards the general certificate for secondary education (GCSE) with its emphasis on a broader range of skills assessed, a lessening of emphasis on the timed exam and an opening up of the exam to a broader section of the age cohort. (All of this was brought in and supported by the same government which is now re-trenching to a formal, exclusive, written exam, but this is another story). The limiting and damaging effect of standardized multiple-choice tests in the United States has also been well documented (Resnick, 1989; Shepard, 1992).

The second question to be asked, and almost never is, is 'what kind of learning do we wish to achieve?' for we know now that different forms of assessment encourage, via their effect on teaching, different styles of learning. If we wish to foster higher order skills including application of knowledge, investigation, analysing, reasoning and interpretation for *all our pupils*, not just the élite, then we need our assessment system to reflect that. But:

> A failure to articulate the relationship between learning and assessment has resulted in a mismatch between the high quality learning described in policy documents as desirable and the poor quality learning that seems likely to result from associated assessment procedures. (Willis, 1992a: 1)

We need to put on to the assessment agenda issues of learning style and depth. We must articulate the model of learning on which we are to base new developments in assessments over the next decade if we are to develop a sound model and one which will achieve the results we wish for it. After all, the original psychometrics was based on a theory of intelligence, and multiple-choice standardized tests were based on a behaviourist model of learning: educational assessment for the next century must be based on current models of learning.

Despite the primacy of assessment to support teaching and learning, assessment for monitoring and accountability purposes will not go away. On the contrary, a number of countries in the developing world are using assessment even more to gear up their education systems; in the United States, in New Zealand, in Australia as in the UK governments have linked economic growth with educational performance and are using assessment to help determine curriculum, and to impose high 'standards' of performance. New Zealand and Britain have taken on board the 'new right' marketplace model, as a market signal to aid parental choice and competition between schools (Murphy, 1990; Willis, 1992b).

This is an awesome role for assessment to fulfil and, mindful now of the distorting effects of assessment for these purposes, the question we must face is how best to design assessment which will provide good quality information about pupils' performance without distorting good teaching (and therefore learning) practice.

The legacy of psychometrics

The science of psychometrics developed from work on intelligence and intelligence testing. The underlying notion was that intelligence was innate and fixed in the way that other inherited characteristics such as skin colour are. Intelligence could therefore be measured (since it was observable like other characteristics) and on the basis of the outcome individuals could be assigned to streams, groups or schools which were appropriate to their intelligence (or 'ability' as it came to be seen). Thus the psychometric testing model is essentially one of limitation: measuring attributes which are a property of the individual and which are fixed. Assessment to support learning, by contrast, aims to help the individual to develop and further his or her learning: it is enabling rather than limiting. This notion of limitation is seen now to be a major disadvantage of the psychometric approach. Linked with this is the interpretation of scores in relation to norms: norm-referencing grades an individual's performance in relation to that of his or her peers, i.e. in terms of relative performance rather than their absolute performance. Norm-referenced tests are designed to produce familiar proportions of high, medium and low scorers. Since students cannot control the performance of other students they cannot control their own grades.

With this model comes an assumption of the primacy of technical issues, notably standardization and reliability. If individuals are to be compared with one another then we need to be certain that the test or assessment was carried out in the same way for all individuals, scored in the same way and the scores interpreted in the same way. Standardization is thus vital in this model as it contributes to the technical reliability of the test. These requirements can have a negative effect on the validity and curricular impact of the test since only some material and certain tasks are amenable to this type of testing.

Along with the psychometric theory and its formulae and quantification comes an aura of objectivity; it is scientific and therefore must be accurate and meaningful. The measurements which individuals amass via such testing, IQ scores, reading ages, ranking, etc. thus come to have a powerful labelling potential.

The psychometric paradigm is based on two other problematic assumptions (Berlak, 1992). First, the *assumption of universality*, which means that a test score has essentially the same meaning for all individuals, i.e. a score on a standardized reading test represents the individual's ability to read (the performance is extrapolated from beyond the test to reading in the general sense) and that what this means is universally accepted and understood.

Second, the notion of *unidimensionality* (Goldstein, 1992, 1993) which relates both to the conceptualization of constructs and to the techniques used for analysing test items. The assumption (within psychometric theory) is that the items in a test should be measuring a single underlying attribute. Items which are 'discrepant', i.e. items which do not correlate highly with the total score, are removed in the test development process because the test is meant to assess only one attribute. Items which have a high correlation with the total score are said to have high 'discrimination' while those which have low correlations are poor discriminators and are usually either dropped or modified.

Since many of the attributes or skills which we measure in tests are multidimensional rather than unidimensional we can see that forcing tests into a unidimensional structure is illogical (Goldstein, 1993) and based on the unproved assumption of unidimensionality. The Rasch model of analysis and the item response model are both predicated on the factor analysis model with a single underlying factor and this is the basis of critiques of these models (see Goldstein, 1992; Goldstein and Wood, 1989).

Tests based on psychometric theory have as a prime requirement good measurement properties amenable to statistical analysis, i.e. reliability and norm-referencing are the prime concerns. As a result of having to meet these requirements, classroom validity or usefulness to teachers have often been overridden or ignored. Around the 1950s the benefit of the application of psychological measurement in educational settings (i.e. educational measurement) began to be questioned and with the publication of *Bloom's Taxonomy of Educational Objectives* educators began to articulate a need for assessment which was specifically for educational purposes and could be used in the cycle of planning, instruction, learning and evaluation.

Educational assessment

Wood (1986) cites Glaser's 1963 paper on criterion-referenced testing (CRT) as a watershed in the development of educational assessment, i.e. the separation of educational assessment from classical psychometrics. Glaser's paper (1963) made the point that the emphasis on norm-referenced testing stemmed from the preoccupation of test theory with aptitude, selection and prediction. Educational assessment, on the other hand, attempts to devise tests which look at the individual as an individual rather than in relation to other individuals and to use measurement constructively to identify strengths and weaknesses individuals might have so as to aid their educational progress.

Wood argues that a powerful reason why educational assessment should not be based on psychometric theory is that the performances or traits being assessed have different properties: 'achievement data arise as a direct result of instruction and are therefore crucially affected by teaching and teachers' (1986: 190). Aptitude and intelligence, by contrast, are traits which are unaffected by such factors. Achievement data are therefore 'dirty' compared with aptitude data and should not/cannot be analysed using models which do not allow for some sort of teaching effect. To find out 'how well' rather then 'how many' requires a quite different approach to test construction. Wood's (1986: 194) definition of educational assessment therefore is that it:

1. deals with the individual's achievement relative to himself rather than to others
2. seeks to test for competence rather than for intelligence
3. takes place in relatively uncontrolled conditions and so does not produce 'well-behaved' data
4. looks for 'best' rather than 'typical' performances
5. is most effective when rules and regulations characteristic of standardized testing are relaxed
6. embodies a constructive outlook on assessment where the aim is to help rather than sentence the individual.

Educational assessment is therefore a model of assessment that can support learning.

Now I turn to models of learning, both traditional and more recent, and what these imply for assessment.

Traditional models of learning

Critics of traditional testing, in particular standardized multiple-choice testing (e.g. Resnick and Resnick, 1992), point out that our testing theory and practice are rooted in a 'traditional' educational model of teaching and testing routine basic skills which developed from psychological theories of learning dating from the earlier part of this century.

'Traditional instructional theory' (based on traditional models of learning) 'assumes that knowledge and skill can be analysed into component parts that function in the same way no matter where they are used' (Resnick, 1989: 3). Psychological theories of the 1920s assumed that learning of complex competencies could be broken down into discrete skills and learnt separately, i.e. through developing individual stimulus-response bonds. This is called the building block model of learning. The idea was that complex skills could be developed later, although the old theory did not make clear how; thus complexity is avoided in the early stages of instruction.

Resnick and Resnick (1992) identify decomposability and decontextualization as two key assumptions of standardized testing technology which are incompatible with developing thinking skills and with what we know now about cognition and learning. The assumption of decomposability comes from the stimulus—response school of learning in the 1920s which posited that knowledge is built of up discrete units or blocks, a collection of independent pieces of knowledge. It thus supports a model of teaching and testing separate skills on the assumption that their 'composition into a complex performance' can be reserved for some time later (Resnick and Resnick, 1992: 42).

Current cognitive theory, however, suggests that this approach is inappropriate. Isolated facts, if learnt, quickly disappear from the memory because they have no meaning and do not fit into the learner's conceptual map. Knowledge learnt in this way is of no use because it cannot be applied, generalized or retrieved. 'Meaning makes learning easier, because the learner knows where to put things in her mental framework, and meaning makes knowledge useful because likely purposes and applications are already part of the understanding' (Shepard, 1992: 319). Skills and knowledge are now understood to be dependent on the context in which they are learnt and practised; facts cannot be learned in isolation and then used in any context. Complex skills are not complex simply because of the number of components involved in them but because of the interactions among the components and the heuristics for calling upon them. Furthermore, assessing separate components will encourage the teaching and practice of isolated components, and this is not sufficient for learning problem solving or thinking skills. '... efforts to assess thinking and problem solving abilities by identifying separate components of those abilities and testing them independently will interfere with effectively teaching such abilities' (Resnick and Resnick, 1992: 43).

The assumption of decontextualization is linked to that of decomposability: 'that each component of a complex skill is fixed, and that it will take the same form no matter where it is used' (p. 43). In fact, what we understand now of cognitive processes indicates that there is an intimate connection between skills and the contexts in which they are used. 'Educationally this suggests that we cannot teach a skill component in one setting and expect it to be applied automatically in another. That means, in turn, that we cannot validly assess a competence in a context very different from the context in which it is practised or used' (Resnick and Resnick, 1992: 43).

Cognitive and constructivist models of learning

An alternative to the linear hierarchy model of learning comes from cognitive and constructivist psychology and shows learning in terms of networks with connections in many directions; not of an external map that is transposed directly into the student's head, but an organic process of re-organizing and restructuring as the student learns.

'Contemporary cognitive psychology has built on the very old idea that things are easier to learn if they make sense' (Shepard, 1991: 8). All learning requires us to make sense of what we are trying to learn and learning through the active construction of mental schemas applies even to young children's 'basic' learning.

So, current cognitive theory suggests that learning is a process of knowledge construction; that learning is knowledge-dependent; and that learning is tuned to the situation in which it takes place. Learning occurs, not by recording information but by interpreting it, i.e. instruction must be seen not as direct transfer of knowledge but as an intervention in an ongoing knowledge construction process. Thus in constructivist learning theory students learn best by actively making sense of new knowledge—making meaning from it and mapping it in to their existing knowledge map/schema; to be useful new information must be linked to the knowledge structures, or schemata, already held in long-term memory. Thus knowledge is seen as something cohesive and holistic which provides a basis for later learning.

This view of student learning which sees students as active constructors of their own world view, including school subject matter, means that we can no longer use an atomistic model of assessment. We need to assess level of understanding and complexity of understanding rather than recognition or regurgitation of facts (Wilson, 1992: 123). 'The strength and frequency of calls for authenticity in assessment are evidence of the influence of such a view of student learning' (p. 123). The basis of most measurement theory is a view of the learner as a passive absorber of information/facts and skills provided by the teacher. Standardized achievement tests test students' abilities to recall and apply facts learnt routinely; even items which are designed to assess higher level activities often require no more than the ability to recall the appropriate formula and to make substitutions to get the correct answer. Many students are succeeding on 'objective' tests without necessarily understanding the material they are learning. But real learning involves constructing one's own interpretations and relating this to existing knowledge and understandings.

One of the concepts that has emerged from work in cognitive science and learning theory is that of *metacognition*. Metacognition is a general term which refers to a second-order form of thinking: thinking about thinking. It includes a variety of self-awareness processes to help plan monitor, orchestrate and control one's own learning. It is a process of being aware of and in control of one's own knowledge and thinking, and therefore learning. Such learners control their learning using particular strategies which hinge on self-questioning in order to get the purpose of learning clear, searching for connections and conflicts with what is already known, creating images, and judging whether understanding of the material is sufficient for the task. An essential aspect of metacognition is that learners control their own learning and, in order to reflect on the meaning of what they are learning, pupils must feel commitment to it.

In the traditional model of teaching the curriculum is seen as a distinct body of information, specified in detail, that can be transmitted to the learner. Assessment here consists of checking whether the information has been received. These newer models of learning, which see learning as a process of personal knowledge construction and meaning making, describe a more complex and diverse process and therefore require assessment to be more diverse and to assess in more depth the structure and quality of students' learning and understanding.

Students who conceive of knowledge as collections of facts will use learning strategies that are aimed at successful memorization. This is influenced by their experience of 'schoolwork' as a required activity which is not directly related to knowledge and learning; their role in traditional schooling is to answer the teachers' questions. The result of this equation of learning with production of work means that learning becomes an incidental rather than an intentional process. A better approach is to develop alternative forms of instruction in which students and teachers jointly engage in knowledge construction and in which teachers progressively turn over metacognitive functions to the students, so that learning is an intentional process and students are taught how to learn. These new conceptions of learning require a new assessment methodology with a focus on tests that facilitate learning and encourage problem solving.

An agenda

In the United States there is a move underway to reduce the amount of standardized multiple choice testing and to replace it with authentic assessment and performance assessment. Performance assessment (PA) is a term widely used in the United States; it is a form of assessment that can be seen in the educational assessment mould. The intention in PA is to

capture in the test task the same demands for critical thinking and knowledge integration as required by the desired criterion performance.
Therefore performance assessments demand that the assessment tasks
themselves are real examples of the skill or learning goals, rather than
proxies. They support good teaching by not requiring teaching to move
away from concepts, higher order skills, in-depth projects, etc. to prepare
for the tests. The focus is more likely to be on thinking to produce an
answer than on eliminating wrong answers as in multiple-choice tests.
Authentic assessment implies that the assessment is authentic to the
learning activity we wish to promote and/or that the context of the
assessment is authentic rather than artificial. Authentic assessment is
always performance-based (it would be difficult to think of an 'authentic'
assessment which was not) so it is in effect a special case of performance
assessment. I shall refer generally to performance assessment. Examples
of performance assessment are practical tasks and written essays (such as
are found in European examinations) and the standard assessment tasks
that were used in the early stages of the national assessment in England
and Wales (DES, 1988). Examples of authentic assessment are portfolios
of students' work.

Another important development, particularly in the UK and Australia,
is the use of teachers' own assessments of pupils' attainment. For example
in the national assessment programme in the UK teacher assessment is a
formal element of the final assessment. In Queensland, Australia, there
have been no subject-based public examinations for 20 years and the bulk
of pupil assessment comes from moderated teacher assessment.

If we are to develop assessment to support learning, and which sits within
the educational assessment paradigm rather than the psychometric
model, we will need to develop assessment programmes which use performance assessment and teacher assessment. Since these are more time
consuming than multiple choice tests or traditional exams, light sampling
needs to be used for national, or local, monitoring programmes. Furthermore, the tasks in which the pupils engage need to be high quality and
pedagogically sound so that the time is not felt to be wasted.

However, when assessments of this type are used for selection or
monitoring purposes, questions about traditional psychometric properties
of reliability and replicability are immediately raised, as are issues of
manageability (Gipps and Stobart, 1993). How can performance assessment be adapted for large scale administration and offer some level of
confidence in comparability of results (which is necessary for accountability
purposes)? There are a number of authors who take the position that, for
these reasons, PA cannot be used for large scale accountability purposes
and of course this highlights the dilemma that we face: there are increased

demands for assessments at national level which must offer comparability, at the same time as our understanding of cognition and learning is telling us that we need assessment to map more directly on to the processes we wish to develop—including higher order skills—which makes achieving such comparability more difficult.

In fact, the GCSE, the English public exam at 16 is an example of a high-stakes, certificating performance-based examination, with a coursework-assessed element. It is taken by the majority of the cohort at age 16; trained markers are used to rate examination scripts outside the school and teacher-assessed coursework is moderated through statistical means and by inspection. Its administration by examination boards and the moderation process means that reliability and comparability are believed to be high. It has, in the view of HMI (1988) and many teachers, had a beneficial, widening effect on teaching and learning. So we have here an example of a performance assessment and teacher assessment examination which is happily used for accountability and certificating purposes.

However, the recent attempt to bring in a performance-based, teacher-assessed national assessment programme at ages 7 and 14 has failed, foundering on issues of manageability and comparability. The lessons from this seem to be first that the system can only cope with such a major assessment programme at one or maybe two ages, and second that if moderation or other audit processes are not carried out then concerns about comparability will cause the public credibility of the assessment programme to fall.

The traditional psychometric notion of reliability needs to be re-thought in these new assessment models (Gipps, 1995). Many of the statistical approaches used in psychometrics do not carry over to assessments with few tasks (or even one, e.g. the portfolio) or those which are scored pass/fail (as with graded assessment, criterion-referenced and standards-based assessment) rather than on a mark scale. We need then to reconceptualize reliability and look instead to other ways of assuring quality in assessment (see also Harlen, 1994). Comparability, across tasks and assessors, is an important alternative. Comparability is vitally important in any assessment used for monitoring, selection, certificating or accountability purposes. Training teachers/raters and moderating their marking and assessment can lead to high levels of comparability across raters. It is also necessary for teacher assessment to train teachers so that they understand the constructs which they are assessing (and therefore what sort of tasks to set); how to get at the pupil's knowledge and understanding (and therefore what sort of questions to ask); and how to elicit the pupil's best performance (which depends on the physical, social and intellectual context in which the assessment takes place).

We also need a more measured, analytical, approach to assessment in education. We need to stop the tendency to think in simplistic terms about one particular form of assessment being better than another: consideration of form without consideration of purpose is wasted effort. We must develop and propagate a wider understanding of its effects on teaching and learning. Assessment does not stand outside teaching and learning but stands in dynamic interaction with it. We must develop a system which supports multiple methods of assessment but at the same time make sure that each one is used appropriately.

Conclusion

My thesis is that assessment for selection, monitoring and accountability can be assessment to support learning. Furthermore, in any testing programme where we test *every* child, it must be so. If we are really concerned about learning for all, and with the child at the centre then we must look at what we know about learning and use that to frame assessment practice. We do have the technology: it is a question of changing our priorities.

We must trust school-based assessment; we must train teachers in observation, diagnostic questioning and formative assessment, give them curricular definitions and exemplars of performance and offer group moderation processes, as well as external audit or moderation. External assessment must be based on a PA model, as in the GCSE. It is expensive, but we need to think about limiting it to one or two ages.

I would argue that training teachers in assessment and supporting school-based assessment needs to be seen as a crucial part of teaching, not as an add on. As for the expense of external assessment you need only look at the figures for how much the World Bank is spending on multiple-choice testing and what the UK government has spent on national testing to see that it is not an issue of resources, it is an issue of paradigm. If we wish assessment to support learning we can make it do so, even when that assessment is to be used for monitoring and selection purposes.

Note

An expanded version of these arguments can be found in Gipps, C. (1995).

References

Berlak, H. (1992). The need for a new science of assessment. In Berlak, H. *et al. Toward a New Science of Educational Testing and Assessment.* Albany: State University of New York.

DES (1988). National Curriculum Task Group on Assessment and Testing: A Report. Department of Education and Science/Welsh Office.

Gipps, C. (1995). *Beyond Testing. Towards a Theory of Educational Assessment.* London: Falmer Press.

Gipps, C. and Stobart, G. (1993). *Assessment: A Teacher's Guide to the Issues*, 2nd edn. London: Hodder & Stoughton.

Glaser, R. (1963). Instructional technology and the measurement of learning outcomes: some questions. *American Psychologist*, **18**(8), 519–521.

Goldstein, H. (1992). *Recontextualising Mental Measurement.* ICRA Research Working Paper. University of London Institute of Education.

Goldstein, H. (1993). Assessing group differences. *Oxford Review of Education*, **19**(2), 141–150.

Goldstein, H. and Wood, R. (1989). Five decades of item response modelling. *British Journal of Mathematical and Statistical Psychology*, **42**, 139–167.

Harlen, W. (ed.) (1994). *Enhancing Quality in Assessment.* Paul Chapman Publishing.

HMI (1979). Aspects of secondary education in England. London: HMSO.

HMI (1988). The introduction of the GCSE in schools 1986–88. London: HMSO.

Murphy, R. (1990). National assessment proposals: analysing the debate. In Flude, M. and Hammer, M. (eds.) *The Education Reform Act 1988.* London: Falmer Press.

Resnick, L. (1989). Introduction. In Resnick, L. (ed.) *Knowing, Learning and Instruction. Essays in Honour of R Glaser.* Hillsdale, NJ: Lawrence Erlbaum Associates.

Resnick, L. B. and Resnick, D. P. (1992). Assessing the thinking curriculum: new tools for educational reform. In Gifford, B. and O'Connor, M. (eds.) *Changing Assessments: Alternative Views of Aptitude, Achievement and Instruction.* London: Kluwer Academic.

Shepard, L. (1991) Psychometricians' beliefs about learning. *Educational Researcher*, **20**(7), 2–16.

Shepard, L. (1992). What policy makers who mandate tests should know about the new psychology of intellectual ability and learning. In Gifford, B. and O'Connor, M. (eds.) *Changing Assessments: Alternative Views of Aptitude, Achievement and Instruction.* London: Kluwer Academic.

Willis, D. (1992a). *Learning and Assessment: Exposing the Inconsistencies of Theory and Practice.* Paper presented at University of London Institute of Education, March 1992.

Willis, D. (1992b). Educational assessment and accountability: a New Zealand case study. *Journal of Educational Policy*, **7**(2), 205–221.

Wilson, M. (1992). Educational leverage from a political necessity: implications of new perspectives on student assessment for Chapter I evaluation. *Educational Evaluation and Policy Analysis*, **14**(2), 123–144.

Wood, R. (1986). The agenda for educational measurement. In Nuttall, D. (ed.) *Assessing Educational Achievement.* London: Falmer Press.

CHAPTER SIXTEEN

Examinations and educational quality

Anthony Somerset

Quality

Examinations, and especially examinations for selection purposes, dominate most young people's experience of education. Teachers and other adults—politicians, employers, journalists—may emphasize other purposes; but to the students themselves it is the final measured outcomes of their education and the opportunities which these lead to, that really matter. Countries differ widely in how high the educational selection stakes have been raised, and in the level at which the most important selection decisions are taken. In some low-income countries, the end of primary schooling is crucial; in a few affluent countries, the major decisions are postponed to as late as the end of a bachelor's degree.

But in the longer run, it is the quality of education provided that is of paramount importance, both for society as a whole, and for students as a part of it. This is why, in discussing examinations, commentators are so sensitive to tension between the role of examinations in selection and in the promotion of learning. As Angela Little points out in her introductory chapter, such conflicts are clearly discernible in most national education systems, and are often the subject of extensive debate. In this chapter I argue that the tension is by no means inevitable, and that, on the contrary, examinations can be used as an instrument to promote quality. To understand their potential, we must start with a model of an entire education system.

Figure 16.1 is an attempt to represent diagrammatically the main components of an education system, and the factors which affect its quality. Like all models of human systems it is an oversimplification; no two-dimensional diagram can adequately represent the subtleties of a major social institution.

The spine of the chart consists of three boxes, outlined in bold: No. 1 (family background); No. 3 (teaching and learning); and No. 14 (life outcomes).

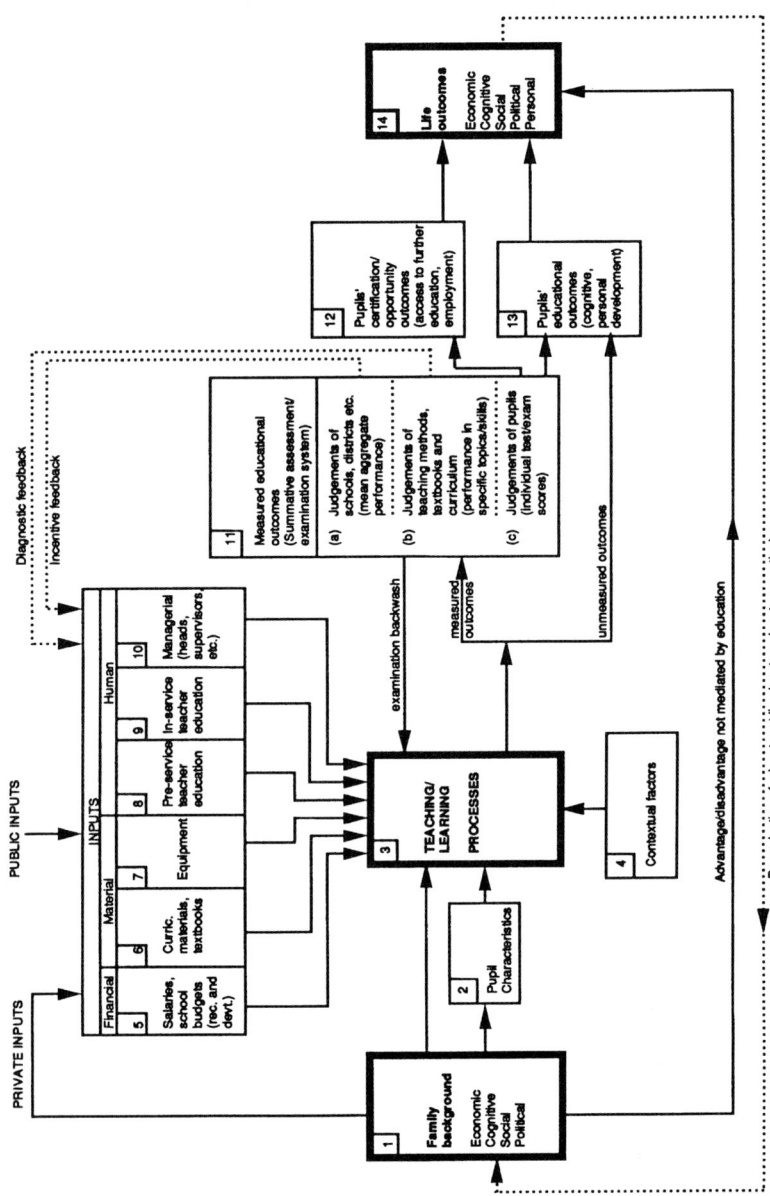

Figure 16.1 A model for educational quality.

Young people are born into families which differ in many ways: in the economic resources at their disposal; in the ways they earn their living; in the intellectual and emotional climate they provide; in the degree to which they wield influence in the community (Box 1).

Ultimately, these children grow up and found families of their own (Box 14). To a large extent, the new families reproduce the patterns of advantage or disadvantage of the original families (direct line linking Box 1 to Box 14). But most societies maintain, to greater or lesser degree, a second channel for élite recruitment: a channel based on 'merit' rather than on family privilege.

The problem, of course, is how to define 'merit'. For a variety of reasons, success in the learning tasks prescribed for the schools has come to be widely accepted as the main criterion. Hence the crucial role of the teaching-learning process (Box 3), and of the education system generally (Boxes 3–11), in mediating the transmission of privilege between the generations.

The role of the formal school as mediator has been exceptionally strong in many newly-independent countries in recent decades. Departing colonial regimes left behind them underformed local élites. For young people who could meet the performance demands of the schools successfully, prospects for economic and social advancement were opened up which had never been accessible to their parents. This capacity of educational systems to promote the upward mobility of a favoured minority was, of course, the main reason for their explosive growth in the post-independence years.

The boxes representing various aspects of the educational system are divided into three main groups: inputs (Boxes 5–10); teaching and learning processes (Box 3); and educational and certification outcomes, usually based on the results of examinations (Boxes 11–13).

In discussing quality issues, various educational stakeholders tend to lay stress on one or other of these aspects, sometimes largely to the exclusion of the other two. Hence there are process definitions, input definitions, and output definitions of educational quality.

Educational quality as process

Professionals with a background in classroom teaching, curriculum development or teacher education often tend to emphasize teaching and learning process as the main criterion of quality. Sometimes it is assumed, without much critical scrutiny, that certain modes of knowledge acquisition, or certain types of interaction between teachers and pupils, or certain forms of classroom organization, are necessarily superior to others. Guided discovery learning, for instance, may be regarded as superior to teacher-directed learning; or learning in groups as superior to whole-class learning; or skills-based learning as superior to content-based learning.

The problem with this approach is that although process is important, it is for the most part a means to an end, rather than the end itself. We want children to enjoy the experience of schooling, of course; and this might be a reason for preferring, for example, guided discovery methods to more didactic methods. Schooling is, after all, life itself as well as preparation for life. But nevertheless, we would not be justified in rating a teacher highly solely on the grounds that she practised guided discovery methods. Before we made a judgement, we would want to know whether her methods led to effective learning. Assessment of process, in short, cannot be divorced from assessment of outcomes.[1]

Educational quality as input

School inputs of various kinds are represented in the figure by Boxes 5–10: financial inputs (Box 5); material inputs, including textbooks and equipment (Boxes 6 and 7); and human inputs, including pedagogical skills, developed through pre-service and in-service teacher education combined with classroom experience (Boxes 8 and 9), and managerial skills (Box 10). The number of boxes is of course arbitrary; each box could be subdivided, or combined with others.

Quantitatively-oriented research workers and development agencies often use input measures to assess educational quality. One major reason is that input data are more readily available than process or performance outcome measures. The number of books in a school's library, the qualifications of its teachers, whether or not it has a science laboratory: these are objective facts which can be collected quickly by relatively unskilled fieldworkers, if they are not already available from a central source.

But the use of input measures as indicators of school quality is even more problematic than the use of process measures. What is significant is not the level of resources going into a school or school system, but rather the manner in which those resources are coordinated and managed for the development of pupil competence. It is common experience in visiting schools to find libraries with books stored in locked cupboards, inaccessible to the pupils; laboratories stocked with equipment but no experiments being carried out; even, in some places, well-qualified teachers listed as staff members, but absent on the day of the visit working in a second job or other income-supplementing activity. Input measures taken alone are poor proxies for quality.

Educational quality as outcome

The four boxes 11–14 represent various outcomes from the educational process. Box 13 is the most general, and the most vague. It represents the sum total of all the competencies which young people acquire, by design or otherwise, through their attendance at school. Cognitive competencies (factual knowledge, concepts, thinking skills), psychomotor competencies (practical skills), and affective competencies (personal development, values, interpersonal skills) are all included.

Some of the outcomes represented by Box 13 are captured by educational measuring systems external to the schools (Box 3 → Box 11 → Box 13), while others remain unmeasured (direct link from Box 3 to Box 13). These external measuring systems are of two main types: examination systems, and assessment systems.

External examinations are a central feature of nearly all education systems (the United States and Sweden being notable exceptions). In countries where examination statistics are publicized, parents, pupils and the general public are much more likely to be impressed by them in making judgements of school quality than they are by inputs or processes. Assessment systems are less common, but likely to become more important in future.

Examination systems and assessment systems. The basic distinction between examinations and assessment systems concerns their purposes and products. The essential purpose of an examination is to differentiate among *pupils*, and its essential product is a list showing how each candidate performed (Box 11c).[2] Hence examinations must generate data at the pupil level: all pupils in the appropriate grade sit the papers.

The essential purpose of assessment systems, on the other hand, is to describe and compare the educational performance of *systems* and *subsystems*, and to monitor changes in their performance over time. Because they are usually based on samples (often very light samples) rather than on full populations, assessment systems have little if anything to say about individual pupils.

The basic product of an assessment system is a set of cross-sectional and longitudinal performance data. The cross-sectional data compare the test performance of various groups at one point in time; the longitudinal data analyse changes in performance over time.[3] If the assessment system is international in scope (for instance, the IEA and IAEP systems: see Chapter 4), there will be comparisons of the performance of the participating countries; but if it is a national system, the comparisons will be between various geographical areas or groups within the country (provinces, rural and urban areas, boys and girls, etc.) (Box 11b).

Performance comparisons can be, and often are, made from the results of an examination. But such comparisons are essentially by-products of the examination, rather than its major focus. Moreover, because examination questions must be changed each year, it is impossible to monitor time-trends in overall performance from an examination. By contrast, the same questions can be used repeatedly in a sample-based assessment system, so time-trend analysis is perfectly feasible.

Examinations

Public examinations are generally held close to the end of an educational cycle; rarely near the middle. They nearly always have consequences for the future prospects of the pupils that sit them (Box 11c → Box 12 → Box 14).

Examinations stakes

It is useful to differentiate examinations according to the weight of these consequences. *Low-stakes* examinations have only minor consequences. The Nepal Primary Leaving (Grade 5) Examination, for instance, is a low-stakes examination, even though it governs access to secondary school. The transition rate from Grades 5 to 6 (the first year of secondary school) was above 85% during the early 1980s, rising to 95% by the middle of the decade. Clearly, passing the examination is a formality rather than a real challenge.

High-stakes examinations, on the other hand, have major consequences. Pupils' performance in high-stakes examinations directly affect their chances of being recruited for the most desirable opportunities in further education, training or employment; and hence can have a substantial impact on their long-term life chances.

The Tanzania Primary Leaving Examination, as it functioned until the early 1980s, probably had higher stakes than any other mass examination in the world, then or since. Fewer than 4% of the candidates were selected for entry to public secondary school each year.[4] No private secondary schools were permitted to function in the country, and there were few other post-primary training opportunities for the 96% who were rejected. The secondary school intake was so small that those who completed the course four years later could be almost certain that a higher secondary place, formal job, or other desirable opportunity would be open to them.

Selection and certification examinations

Public examinations are often classified into two types, according to the main purpose for which they are run: selection examinations, which are specifically designed to identify recruits for a particular type of opportunity—most often, places in the next cycle of formal education; and certification examinations, which provide candidates with a record of achievement in the cycle they have just completed. The distinction is far from clear-cut—many selection examinations have certification functions, and most certification examinations have selection functions—but is nevertheless useful.

Table 16.1 summarizes some contrasts between selection and certification examinations, all stemming from the central difference in purpose. It should be stressed that the table compares ideal types: many if not most public examinations are intermediate in character.

The table is self-explanatory, but one point, concerning the criteria of 'success' in certification examinations, perhaps requires elaboration. Partly out of concern to avoid stigmatizing large numbers of young people as failures, partly for broader social and political reasons, examinations authorities often set low minimum standards for the award of a formal pass in certification examinations. In the Sri Lanka General Certificate of Education (GCE) examination, for example, a raw mark of only about 35% secures a pass in an individual subject. In most years, at least 70–75% of candidates pass in six or more subjects, and thus qualify for a GCE certificate (University of Cambridge Local Examinations Syndicate, 1990: 60).[5]

But in nearly all certification examinations, passes are differentiated, often into a number of ranked grades; and much higher achievement standards are set for the upper grades. In Sri Lanka, a raw mark of 50% is needed for a credit pass, and 75% for a distinction.[6]

In most countries, the opportunity market which certificate holders enter is highly competitive, so that lower-level 'pass' grades often do little or nothing to improve prospects. If their search for opportunities proves fruitless, candidates with lower-level passes are likely to regard themselves as failures—despite what is recorded on their certificates.

The Indonesia University Entrance Examination (Ujian Masuk Pendidikan Tinggi, or UMPT) is an example of a pure selection examination. UMPT is run entirely by the public universities, through *ad hoc* committees made up of academic faculty members. The only results issued are lists of applicants accepted to the various courses. Candidates are not told their grades, so those who fail to gain a university place (about 90% of all candidates) have no certificate recording their achievement which they might use to seek other opportunities. The whole process is completed with great rapidity: the period of time which elapses between candidates sitting the examination papers and recruits starting their university courses is less than three months.

Table 16.1 Selection examinations and certification examinations

	Selection examinations	**Certification examinations**
Access to subsequent opportunities	Access direct and usually rapid for successful candidates. Typically, opportunities offered by the recruiters: candidates do not actively seek them	Access relatively indirect. Candidates must actively seek opportunities. Search often prolonged; may well be fruitless
Range of subsequent opportunities	Generally only a single type of opportunity available; most often secondary school or university places	Broader range of opportunities; likely to include employment and/or preservice training
Criteria for recruitment	Examination performance the main, often the sole, criterion for recruitment	Examination performance usually not the sole criterion for recruitment[7]
Certification	Examination authority may or may not issue a certificate. If it does, likely to be useful simply as a record of achievement (not as 'currency')	Authority issues a certificate indicating performance, which the candidate then uses as 'currency' in his/her search for opportunities. Value of certificate depends on grades
Criteria of 'success' in examination	Narrow and clearcut: gaining a place constitutes 'success'; not gaining a place constitutes 'failure'	More ambiguous. Proportion who formally 'pass' often high, but candidates with lower-grade passes likely to regard themselves as failures if search for opportunities proves fruitless
Control of examination	Recruiters usually. Influential university selection examinations sometimes run entirely by universities, with little or no input from other stakeholders	Often a broader representation of stakeholder interests—especially the interests of those responsible for preparing candidates—than in control of selection examinations

Indonesia also provides an example of a certification examination with virtually no selection functions. The Ebtanas examination, administered by the Ministry of Education, operates in the shadow of the UMPT at the transition point between secondary school and university. An Ebtanas pass is needed for sitting the university entrance examination, but in practice this requirement is little more than a formality, because the pass-rate is well above 90%. In effect, the Ebtanas certificate is a consolation prize for those who fail to gain a public university place. It provides a record of secondary school graduation (which UMPT does not), but has little value as currency in the labour market. Few formal jobs are available, so most secondary school leavers who are not selected through UMPT enrol in private universities if they can afford the fees.[8]

As already mentioned, however, examinations intermediate in character between selection and certification examinations are perhaps the commonest type, especially at the junction between secondary school and university. These examinations are often selection examinations for university recruitment, but certification examinations for other opportunities. The highest-scoring candidates are recruited directly by the universities, whereas lower-scoring candidates must seek for opportunities themselves, using their certificates as 'currency'.

The names of public examinations tend to stress their role as certification instruments rather than as selection instruments: 'Certificate of Primary Education', 'School Leaving Examination' and 'School Certificate' are commoner names than 'Secondary Selection Examination' or 'University Entrance Examination'. But whatever their formal title, most public examinations do, in fact, have explicit or implicit selection functions. And it is these functions which make them so important.

Examination tiering. The number of public examinations in the school system varies from country to country. Practically all countries (the Unites States and Sweden excepted) have an examination (or series of examinations) at the end of the secondary cycle, regulating access to university. In the Asia–Pacific region, two-tier systems are common; the university entrance examination being preceded by an examination regulating the transition between lower secondary and upper secondary school. In the Africa region many countries maintain a full three-tier system: examinations are held at the end of each school cycle (primary, lower secondary, upper secondary), each governing access to the next cycle.

Examination output

Examinations can generate three main types of information, represented in the model by sections 11a, 11b, and 11c.

Judgements of pupils. Box 11c represents the traditional, and essential, product of examinations: judgements of individual pupils, based on aggregate achievement, and perhaps the profile of achievement over the range of subjects tested. These judgements are, of course, the basis on which educational certificates are issued and selection decisions made (Box 11c → Box 12).

Judgements of schools and districts. Box 11a represents judgements of schools, districts, and perhaps other units of the education system which can be based on examination results. School- and district-level achievement means can be calculated, and used for accountability purposes (dotted line from Box 11a to input boxes: 'incentive feedback'). Depending on how high the examination stakes are, and on how widely the results are publicized, teachers and head teachers in schools where results are poor are likely to come under considerable pressure from educational managers, parents, politicians, and the public generally to improve achievement. Similarly, if the weak schools tend to be concentrated in particular districts, regions or other administrative units, then the performance of the educational managers responsible for those units is also likely to come under scrutiny.

Most teachers, heads, and educational managers are, of course, professionally concerned to ensure that the pupils in their charge perform well in public examinations; but the introduction of performance monitoring means that they have a personal stake as well. An examination which is high-stakes for the candidates becomes high-stakes for those responsible for instructional management also. Hence it is important to ensure that the monitoring system makes allowance for factors which might affect scores but for which instructional managers cannot be held accountable. Such factors include differences in family background (Box 1) and resource inputs (Boxes 5–10). It would, for example, be grossly unfair to expect teachers working in the deprived slums of a major city to match the results achieved by schools in the high-income suburbs.

In low-income countries with a high proportion of rural schools, however, differences between catchment areas may present less of a problem. As Heyneman's studies first showed, educational performance is less affected by socio-economic and other out-of-school factors in such countries than it is in more developed, more urbanized countries.

Kenya's experience in introducing a performance feedback system is instructive. District-level and school-level mean scores for the national primary leaving examination (the Certificate of Primary Education) were first calculated after the 1976 sitting. It was found that, although in general urban districts performed better than rural districts, there were a number of exceptions. Nyeri, the top rural district, was second in the overall ranking, out-performing Nairobi city by nearly half a standard deviation.

Mombasa, the bottom urban district, ranked as low as 31st among 43 districts (39 rural, four urban). Within districts, there were numerous instances of schools in remote areas standing higher on the performance lists than well-established schools close to the district centres. It was therefore decided to make both the district-level and the school-level performance lists public, but with separate rankings for the urban and rural districts.

The effects were dramatic: the quality of primary education immediately became a matter of national concern. Public meetings to discuss examination performance were held in at least ten districts.[9] Actions to improve matters varied: in some districts, trophies for the most successful schools were established; in others, parent-teacher associations formed. In one district (Murang'a), more fundamental changes in the school management and support system were instituted. A number of education officers and inspectors were replaced, as were the heads of some of the weakest schools. In 1976, when performance means were first calculated, Murang'a was among the five weakest districts; by 1983, seven years later, it had displaced Nyeri as top rural district, and was second among all districts, rural and urban.[10] An equally impressive improvement occurred a few years later in Turkana, an arid, pastoral district in one of the remotest and least-developed corners of the Republic. In 1979 and 1980, Turkana ranked among the bottom three rural districts; a decade later, it was regularly among the top three.

Judgements of teaching and learning. Finally, Box 11b represents judgements which can be made of teaching and learning, and sometimes textbooks and curriculum, through analysis of performance by topics, cognitive skills, and individual questions. Instead of totalling scores horizontally, across questions, for each pupil or group of pupils, they are totalled vertically, across pupils, for each question. Furthermore, the pupils' responses are not recorded simply as 'right' or 'wrong'; the pattern of the various incorrect responses to each question is also analysed. This pattern can be the basis of a diagnostic feedback system, providing school heads, teachers, advisers, teacher educators, curriculum developers, and textbook writers with insights into the reasons why pupils regularly make particular errors, and the steps which could be taken to help eliminate such errors (dotted line from Box 11a to the input boxes: 'diagnostic feedback').

Analysis of error patterns is, of course, greatly facilitated if the examination questions are in multiple-choice format. Nevertheless it is entirely practicable to analyse open-ended responses in the same way, by using light samples of the scripts.

The value of the information is considerably enhanced if the error patterns for major groups are compared: rural schools with urban schools; élite schools with non-élite schools; boys with girls, for example.

It should be stressed that public examinations are not suitable instruments for the fine-grained analysis of learning difficulties in particular topic or concept areas. For these purposes, specially-constructed tests are needed. Examinations must aim for broad coverage of the entire curriculum, and so cannot focus on particular topics in great detail. Nor are they appropriate for diagnosing the strengths and weaknesses of individual pupils, because they take place at the end of the teaching cycle, when remedial action is no longer possible.

Nevertheless, examinations can provide a wide range of insights into the learning difficulties shared by large groups of pupils. The following brief examples are abstracted from feedback booklets prepared for Kenya primary teachers, based on analyses of performance patterns in the Kenya Certificate of Primary Education (KCPE) examination.

1. *Time problems.* In the analysis of answer patterns to a multiple-choice question involving a time calculation included in the 1979 mathematics paper, it was found that about one-third of candidates chose an answer which indicated they were working in base 10 instead of base 60 when adding the minutes. They added 2.55 hours to 10.50 p.m. to get 1.05 a.m. as their answer, not realizing that the point in time notation is not a decimal point, but simply a marker separating the hours from the minutes. A further 28% gave as their answer 1.45 p.m., instead of 1.45 a.m., suggesting that they have a poor understanding of the a.m./p.m. system for recording time. (The Newsletter continues with several pages of suggestions as to how teachers can develop pupils' skills in solving time problems). (1980 Newsletter)
2. *Girls and science.* In the 1981 science paper, girls performed poorly, relative to boys, on questions which required them to recall the results of observations and experiments, but much better when required to carry out an experiment during the examination. They also performed relatively well when asked to reason on the basis of given information. 'It seems that girls can do experiments, given the chance, but do not participate fully during the learning process.' 'Make sure that when you are carrying out practical work in science, the girls take just as active a part in the observations and experiments as the boys.' 'Girls should sometimes work separately from boys. This may either involve forming separate groups or each child may work individually.' (1980 and 1982 Newsletters)

Extended extracts from Kenya CPE Newsletters, giving numerous examples of diagnostic question analysis, are available in Somerset (1987: 71–72, 120–151, 1988: 183–186).

Obtaining the raw statistical data for a diagnostic feedback system is straightforward, especially if the questions are in multiple-choice format and have been marked by computer. But identifying the most significant trends in the answer patterns, interpreting the reasons for them, and writing about them in ways that are meaningful to classroom teachers, is highly demanding work, requiring experience, skill, and time. Very often, it is necessary to visit schools and discuss the error patterns with teachers and pupils before writing the feedback document.

Major national examinations are, of course, basically summative instruments. But this does not preclude their having formative functions as well. While judging the achievements of pupils who have just completed the schooling cycle they can, at the same time, help strengthen the achievements of pupils who are following after. Few countries, however, have as yet begun to tap their considerable potential.

Examination backwash

It is a common complaint, to be heard in virtually every country where there is a high-stakes examination, that 'teachers teach to the test'. External examinations send signals back to the schools as to the topics and skills they should concentrate on, and the instructional approaches they should adopt (Box 11 → Box 3: examination backwash'). The scarcer, and more valued, the opportunities allocated on the basis of the examination, the stronger the backwash effects will be. So there is often substantial circularity in the relationship between teaching/learning processes and examinations: the examinations measure the outcomes of the processes (Box 3 → Box 11); but they are also a powerful determinant of them (Box 11 → Box 3). High quality examinations, by sending appropriate signals to the schools, help promote high-quality instruction; whereas low-quality examinations, by sending inappropriate signals, promote low-quality instruction.

Backwash effects are strongest, of course, in the period when pupils are being actively prepared for the examination. Depending on how high the stakes are, this period may last for a semester, a year, or several years. But the shadow of the coming examination is often apparent throughout the preceding school cycle. Teachers in the lower classes of the cycle frequently model their internal tests and examinations on the format employed for the external examination, even if it is several years away and the content they are testing is quite different.

Of course no set of examination questions, however skilfully devised, can do more than sample the broad range of competencies which a well-designed curriculum will aim to develop. An examination is over in a few hours or a few days; whereas a curriculum may take several years to deliver. The real issue is whether or not the sample of competencies tapped by the examination is representative of the curriculum. Too often, it is not.

In fact, the disjunction between the competencies which a country seeks to develop in its young people through its curricula and the competencies measured by its national examinations is often substantial. There are two main ways in which examinations in many countries fail to reflect curricula.

1. *Examinations often fail to take sufficient account of the cognitive needs of terminating students.* High-stakes examinations are very often biased towards the testing of competencies needed by the students continuing their education into the next cycle (usually the minority); and against the testing of competencies especially relevant to terminating students. In mathematics and science especially, questions concerned with the more specialized and abstract topics tend to predominate heavily over topics with more practical application in the everyday world.
2. *Examinations often depend too heavily on recall.* In many examinations, a high proportion of questions requires the candidate simply to reproduce learned material directly from memory, without reconstructing it or using it in any way. Very often the tested material consists largely of 'factoids'—isolated fragments such as names, dates, places, technical terms—the building blocks of knowledge rather than its structures.

Such examinations embody a view of pedagogy as transmission rather than as construction. Students are seen, essentially, as receptacles. The job of the teacher is to fill them with knowledge, and the job of the examination to measure how much of that knowledge they have retained. The need for students to be active participants in the building of their own knowledge is not recognized. Hence few questions are asked which test pupils' capacity to apply what they have learned to new situations, or which require them to show they understand how facts link together, in meaningful patterns of cause and effect. Similarly, relatively little attention tends to be paid to the testing of higher-order skills: the ability to assimilate and use new information; the ability to infer conclusions from new or remembered information; the ability to develop a logical sequence of steps to solve a problem or reach a decision; the ability to argue a case or present a point of view; the ability to produce creative or imaginative work.

There are two main reasons why the quality of major external examinations is often so low: first, institutional isolation; and second, time and resource pressures.

1. *Institutional isolation.* In many countries, examinations managers maintain a protective wall around their operations. This is done partly for security reasons—the smaller the number of people entrusted with the preparation and review of examination papers, the lower the risk of leakages—but also, very often, because the managers see it as being

important not to become too closely identified with any of the numerous groups with a stake in the examinations results: universities, employers, ministry officials, teachers, parents, politicians. Within the wall, an examination culture builds up, so papers become stereotyped and predictable. But in some cases the wall is constructed from both sides: 'With good justification, professional educators ... tend to have ambivalent feelings about selection examinations ... They see [them] as being at best a necessary evil, to be tolerated but not encouraged; at worst, as a barrier to progress which should be swept away. They prefer to devote their energies to work which they regard as more creative: the development of new curricula, the in-servicing of teachers, the preparation of new teaching materials' (Somerset, 1987: 15).

2. *Time and resource pressures*. The annual cycle of activities involved in running a major examination system subjects examinations managers in many developing countries to unrelenting pressure. Numerous deadlines must be met; and, moreover, many of these deadlines are linked sequentially, so that failure to meet an 'upstream' deadline can have severe knock-on effects on 'downstream' activities. Substantial increases in candidate numbers may have to be accommodated without a corresponding increase in staffing.

For most examining authorities, the prospect of administrative failure is more threatening than the prospect of professional failure. Public interest in examinations is intense, but in most countries it focuses on their allocational rather than their educational functions. Many people will wish to see the results lists—which candidates have the highest marks, or have been selected for the next cycle, which schools or districts have performed best—but relatively few will be concerned with the quality of the measuring instruments on which the results are based.

Hence, if an examinations organization is under stress, it will tend to focus its efforts on the achievement of administrative goals, at the expense of professional goals. If a choice must be made, it is more important to ensure that the examination runs smoothly, that there are no leakages, and that results are issued on time than it is to ensure that the results provide a valid measure of an appropriate mix of cognitive competencies.

Whether because of institutional isolation, time and resource pressures, or a combination of both, the quality of the examination often suffers. Too few people take part in the writing of questions, and the same few tend to be involved from year to year. Too little time is devoted to reviewing the questions, and to the preparation of balanced papers. As a consequence, questions testing recall of factual material tend to predominate heavily over questions testing higher-order skills, because recall questions are much quicker, and much easier, to write. Identical or near-identical questions tend to be repeated frequently. Similarly, the range of formats used in the examination is likely to be restricted to those which create the fewest administrative problems, regardless of professional considerations. Straightforward paper-and-pencil tests, for example, will almost certainly be strongly emphasized; whereas practical tests, including oral tests in language subjects and tests of observational and experimental work in the sciences, may be absent from the examination altogether—no matter how important the role allocated to the development of practical skills in the formal curriculum.

The backwash signals which such examinations send to the schools are clear and strong: first, that pupils who wish to achieve high scores should concentrate on memorizing factual material, particularly material frequently included in previous examinations; and second, that time devoted to the development of information processing, problem solving or practical skills is likely to be time wasted.

Dealing with examinations backwash

Heyneman has summarized succinctly the three choices open to educational managers confronted with examination backwash: 'They can fight it; they can ignore it; or they can use it' (Heyneman and Fägerlind, 1988).

Fighting backwash

1. *The Dore proposals.* In his celebrated analysis of the 'diploma disease', Dore (1976) makes several suggestions as to how the negative backwash effects of examinations on learning in the schools might be combated. All of them involve reducing the power of examinations to influence life outcomes. Dore proposes that within the school system, and at the junction between schooling and work, the use of learning achievement tests for selection should be avoided. Instead, aptitude tests, or other 'tests which cannot be (or cannot much be) crammed for' should be employed. In some circumstances, Dore suggests, lotteries might be appropriate, especially after preliminary screening of applicants on the basis of aptitude test scores. Employment might be started earlier, after a basic education cycle open to all, with extended opportunities for careers development through on-the-job training and short-term full-time courses (including university-level courses). Selection into the higher-level occupations would thus take place at the workplace, and be based on job performance rather than examination scores.

Little (1984) has provided a careful critique of these proposals. As she points out, aptitude tests are far from immune to the effects of instruction. When verbal reasoning, numerical reasoning, or other aptitude-type questions are included in high-stakes examinations, teachers nearly always devote considerable attention to training their students in how to tackle them.[11] Contrary to common opinion, 'aptitude' tests do not measure innate capacities or underdeveloped potential; it would be more accurate, if more cumbersome, to style them curriculum-reduced tests of developed abilities.

Lotteries have been used to a limited degree for selection, usually in situations where there are large numbers of applicants with near-equivalent qualifications. But they are unlikely to ever gain general acceptance because they are widely seen as contradicting the meritocratic ideal: worth, not luck, should decide who gets the best life chances.

Earlier selection for employment, with extended opportunities for on-the-job training and qualification upgrading, would lead to much closer linkages between learning and work; but labour costs for employers would be higher, and the reform would meet resistance from stakeholders in the present system of higher-secondary and post-secondary education.

2. *Internal assessment*. An alternative way to reduce the power of external examinations to affect instruction, less radical than those proposed by Dore, would be to transfer some of the responsibility for assessing school leavers to the schools themselves. Internal assessment has been introduced in a number of countries, most often to complement, but in a few cases to replace altogether, external assessment through an examination.

Internal assessment has some obvious advantages. Because measurements can be made over an extended period rather than at a single time, the high levels of anxiety often provoked by external examinations are largely avoided. For the same reason, a more rounded view of each pupil's strengths and weaknesses can be built up. Trends in performance can be identified, and allowances made if performance at a particular time is affected by ill-health, personal or family stress, or bad luck.

Nevertheless, internal assessment has its problems—different for the most part from the problems of external examinations, but no easier to solve. We shall discuss three of the most important.

(a) *The quality of teacher-made tests*. The accuracy and fairness of any assessment system depends on the quality of the instruments used. We have already noted that in many countries the question papers set for external examinations show major weaknesses: they depend too heavily on recall, particularly on the recall of fragmented factual material; they pay too little attention to information-processing and problem-solving skills; they tend to focus on abstract and specialized topics rather than on topics with practical application. When national examinations are defective, teacher-made tests nearly always suffer from the same weaknesses, but often to an even more marked degree. For internal assessment to work well, substantial resources must be devoted to the development of teachers' assessment skills.

(b) *The need for a common yardstick.* Competent and committed teachers can rank the pupils in their charge with considerable accuracy. But with the best will in the world, they cannot judge how their pupils compare with pupils in other schools, where they do not teach. To do this, they need a common yardstick.[12]

Various possibilities are available. If the examination has both internally and externally assessed components, the simplest method is to use the external component as the yardstick. This was the approach adopted by Tanzania, when it introduced internal assessment into its secondary school leaving examination in the mid-1970s. For each subject and each school, the distribution of grades awarded by the teachers was compared with the same distribution from the external examination. Where discrepancies were found, corrections were applied.

A few high-income countries, with well-developed educational infrastructures, have set up systems employing professional judgement as well as statistical data for monitoring and coordinating schools' grading standards. In New Zealand, where a system of school-based accreditation for university entrance was introduced as long ago as 1946, standards were monitored by school inspectors. Because the number of secondary schools was small, the inspectors could build up a good deal of first-hand knowledge as to the strengths and weaknesses of each. But monitoring was informed by statistical evidence as well as professional judgement. For students who were not accredited, the traditional university entrance examination was retained. Scores in this examination provided an efficient means to identify schools where accreditation standards were too lenient, or too severe (Parkyn, 1958).

In Sweden, by contrast, the external university entrance examination was abolished altogether when internal assessment was introduced during the 1970s. Instead, scores on standardized achievement tests, set nationally but administered by the appropriate subject teacher, are used to monitor standards. Discrepancies between test scores and grades based on cumulative records of achievement are discussed between the school principal and the subject teachers, and further meetings are held to establish comparability among schools. The grades finally agreed form the basis for university selection (Marklund, 1988).[13]

Internal assessment can play an invaluable role in a total system of outcome evaluation, but only if a criterion is available for measuring quality differences among schools, and controlling their effects. For most low-income countries, lacking the resources to set up a monitoring system such as the one developed in Sweden, the most effective criterion is likely to be an external examination.[14]

(c) *Internal assessment and the teacher's role.* The introduction of internal assessment into a national examination system brings about major changes in the role of teachers, particularly where the examination stakes are high. When teachers take no part in the examination, their interests and those of their pupils coincide: 'students and teachers can become allies in the business of maximizing examination performance' (Ekstein and Noah, 1993). But when teachers become examiners, their role in the classroom becomes ambiguous. On the one hand, they are required to promote and encourage the cognitive growth of each of their pupils; on the other, they are required to make judgements as to how much growth each pupil has achieved—judgements which may have profound consequences for their future life chances.

Clearly, there is tension between these two sets of requirements. The teacher's power to influence her pupils' futures may well have negative effects on her power to promote their current learning. Pupils may, for instance, be inhibited from asking questions, for fear that questioning may be taken as evidence of lack of understanding or slow progress. Similarly, pupils who perform poorly early in the course may give up trying, because they feel that the teacher's mind is already make up. Skilful teaching can reduce or even eliminate these effects, but it is doubtful that the underlying tension can ever be fully resolved.

The teacher's power to influence her pupils' futures may also have consequences outside the classroom. The case of Tanzania is instructive. As we have already noted, Tanzania successfully introduced an internal assessment component into its secondary school leaving examination in the 1970s. But a parallel initiative with the primary leaving examination failed. As in most countries, Tanzania secondary teachers are subject specialists; so internal assessments are collective judgements, to which many teachers contribute. Moreover, most secondary schools are large, and a high proportion of pupils are boarders. Hence face-to-face contacts between teachers and parents are infrequent. At the primary level, by contrast, most primary schools are small, and most teachers generalists. Hence the group with responsibility for internal assessment may consist of as few as two or three people. Furthermore, most primary schools serve small, local catchment areas, so teachers have frequent contacts with parents—particularly with those who are influential local leaders. In these circumstances, to require teachers to make judgements which might profoundly affect their pupils' life chances would be to place a burden on them which few human beings would wish to carry.

Ignoring backwash

This is the commonest, but the most dangerous, alternative. The backwash effects of a high-stakes examination are too powerful to be ignored. The point made at the beginning of this section concerning the essential circularity of the relationship between teaching and testing is crucial: a high-stakes examination is a powerful input to the quality of teaching and learning, as well as a measure of it.

The most dedicated efforts to improve instructional quality through teacher inservicing, curriculum and textbook improvement, and other means can be sabotaged by a badly-set high-stakes examination. It is unrealistic, as well as unfair, to ask teachers to adopt new pedagogical approaches if the main criterion against which their competence and success is judged remains unreformed. *Low-quality examinations promote low-quality education.*

Using backwash

In nearly all contexts the third alternative, to use examinations backwash constructively, rather than to fight it or ignore it, is most appropriate. There is no intrinsic reason why the backwash effects of examinations on teaching and learning should be harmful; they are harmful only when the examinations are of inadequate quality. If examinations were to receive the same resources of professional skill as are devoted to, for instance, curriculum materials development and teacher education, they could become a spearhead of educational improvement, rather than an obstacle to it.

Notes

This chapter is based on an extract from a longer document prepared for the Asia Technical Department of the World Bank entitled *Examinations and the Quality of Education. Issues for the Asia Region* in 1993. Permission from the Asia Technical Department to publish the material in this form is gratefully acknowledged.

1. Overemphasis on process, at the expense of outcomes, is a frequent weakness of teaching practice assessment during initial teacher education. Because classroom observation is much more labour-intensive than most other modes of assessment, teacher educators tend to rate trainees on aspects of practice which are readily apparent during a short visit: the preparation of lesson plans,

classroom management, use of teaching aids, etc. But evidence as to whether or not the trainee's approaches are leading to effective learning takes longer to gather, and so tends to be overlooked.

2. Although the data released to candidates may be limited. In the Indonesia university entrance examination, for example, three-digit standard scores are calculated for each candidate, allowing very fine differentiation in the selection process; but the only results made public are lists of candidates whose applications to the various university programmes were successful.

3. If the assessment system is newly established and only one round of testing has been completed there will, of course, be cross-sectional data only.

4. In 1979 the proportions selected were: 4.2% of boys, 2.8% of girls.

5. The tying of grade boundaries to fixed raw marks means, of course, that no allowance can be made for variations in the difficulty of the papers from year to year.

6. Minimum standards are even more generous in the recently-established General Certificate of Secondary Education (GCSE) in England and Wales. The terms 'pass' and 'fail' are avoided; instead, candidates receive an 'award', in one of seven levels, A–G. In most subjects, fewer than 5% of candidates receive no award at all. But in the popular perception, the bottom three or four levels are widely regarded as failing grades, because they do not as a rule qualify the candidate to enter a higher secondary ('Advanced-level') course.

7. Employers vary a great deal in the weighting they give to examination results when making recruitment decisions. They often, however, set a minimum examination grade as a preliminary filter, to limit the number of applicants to be considered.

8. There are more than 1000 private institutions providing post-secondary pro-grammes at the degree level, as compared with only 45 public institutions. Private universities are of two quite different types: a handful of élite institutions, offering high-cost education of a quality comparable with the best public universities, and huge numbers of second-choice institutions of mediocre or poor quality, which recruit students who failed to qualify for the public universities.

9. According to newspaper reports. It is likely that meetings in other districts went unreported.

10. This transformation was celebrated at the time as the 'Murang'a Miracle'. Murang'a and Nyeri are adjacent districts and traditional rivals, which doubt-less contributed to Murang'a's rapid improvement once the existence of the performance gap became known. See Somerset (1988: 100–107), for further details of the effects of the dissemination of examination performance data in Kenya.

11. And this preparation brings results. In one rural district of Kenya, it was found in the English paper of the secondary school selection examination that the performance gap between high-scoring and low-scoring schools, closely matched in socio-economic characteristics, was just as wide in verbal reason-ing items as in achievement items. (Somerset, 1987).

12. In a study carried out in Indonesia comparing the results of internal and external assessment in the selection system for entry to university, it was found

that even the schools which were most efficient at ranking their students in performance order (as judged by the correlations between internal and external scores) were quite unable to make valid judgements as to the mean score that should be assigned their students. In fact, several of the top schools (as judged by the external examination) awarded lower mean scores to their own candidates than many of the weaker schools—presumably because they set higher attainment targets.

13. In 1991, however, a second pathway for university recruitment was established: students were given the option of qualifying through sitting a national Scholastic Aptitude Test, rather than through the school-based assessment system. This suggests that the school-based system 'may not be working to produce a wholly satisfactory degree of comparability' (Eckstein and Noah, 1993).

14. In the United States, where in most states there are no formal external examinations, university recruiters often use scores on the Scholastic Aptitude Test (SAT) to measure differences in assessment standards among schools.

References

Dore, R. P. (1976). *The Diploma Disease*. London: Allen & Unwin.

Eckstein, M. A. and Noah, H. J. (1993). *A Comparative Study of Secondary School Examinations*. Research Working Paper No. 7, International Centre for Research on Assessment, Institute of Education, University of London.

Heyneman, S. P. and Fägerlind, I. (eds.) (1988). *University Examinations and Standardised Testing*. Technical Paper No. 78, World Bank, Washington, DC.

Little, A. (1984). Combating the Diploma Disease. Chapter 7 in Oxenham, John, *Education versus Qualifications?* London: Unwin.

Marklund, S. (1988). Education in Sweden: Assessment of student achievement and selection for higher education. In Heyneman and Fägerlind (eds.) *op cit.*

Parkyn, G.W. (1959). *Success and Failure at the University. Vol. 1: Academic Performance and the Entrance Standard*. Wellington: New Zealand Council for Educational Research.

Somerset, A. (1987). *Examinations Reform: the Kenya Experience*. Report No. EDT64, World Bank, Washington, DC.

Somerset, A. (1988). Examinations as an instrument to improve pedagogy. In Heyneman and Fägerlind (eds.) *op cit.*

University of Cambridge Local Examinations Syndicate (1990). *Educational Assessment in Sri Lanka: A Report on the Education and Examinations Systems in Sri Lanka*. Cambridge, UK: UCLES.

CHAPTER SEVENTEEN

Individual choices, incentives and control: understanding assessment dilemmas

Alison Wolf

One of the themes of this book has been the difficulty of achieving assessment reform, and in particular the way in which reforms founder because of the competing pressures on teachers, or the difficulty of breaking away from the tyranny of academic selection. In Chapter 1, for example, Angela Little refers to the way in which attempts to broaden examination content in Malaysia, and to introduce school-based assessment in Sri Lanka, both failed: and other chapters—for example those on Egypt, England and Wales and the United States of America—also describe major reform efforts which failed to meet their original objectives.

There are no assured ways of avoiding such failures. They reflect very real dilemmas which are at best difficult and may often be impossible to resolve. However, this chapter argues that, if assessment reforms are to be more successful in the future, we need to make more use of a theoretical paradigm which is not much in evidence in current writing on assessment. We need to think in 'economic' terms: about pupils' behaviour and also about that of teachers and administrators.

It may seem strange to argue that assessment studies, or any other parts of modern educational research, ignore economics, given the predominance of arguments about the contribution of education (and training) to national wealth. The idea that education should be about something more than 'utility'—that it should be about a cultivated mind, and concerned with 'liberal knowledge which stands on its own pretensions, which is independent of sequel' (Newman, 1852) is heard increasingly rarely: and, when it is, is advanced defensively, by authors who are ready for attacks on irrelevance and ivory towers. Instead, one finds an overwhelming (if largely unsubstantiated) consensus that more education and training are the key to economic success. It is this consensus which lies behind so many governments' current interest in the use of assessment for monitoring, for improving educational quality, and in the development of new, more elaborate qualification systems designed to provide the economy with skilled personnel.

However, in arguing the need for more use of an economic perspective, this is not what I have in mind. The fundamental economics paradigm focuses not on governments or 'national interest', but instead on the individual. The assumptions and propositions that make up this paradigm have to do with the *rationality of individual behaviour*, and they are the building blocks of all classical economic theory, including that of theorists as different as Karl Marx and Milton Friedman. In recent years, other social sciences, particularly political science but also sociology and anthropology, have adapted the economic paradigm in developing theories of '*rational choice*'; the term 'choice' underlining the fact that individuals are constantly obliged to decide between alternative courses of action.

Many of the chapters in this book emphasize the labour market pressures facing individual *students*, and their resulting tendency to work for ever-more qualifications, and to focus narrowly on material which is relevant to formal examination success. This chapter suggests that we need to look not just at the choices and pressures facing learners but to consider other 'players' in the assessment game in the same way. Many different individuals are involved in any assessment system, as students and test-takers, but also as assessors, teachers, administrators or paymasters. Assessment research should pay more attention to individual behaviour generally: in particular to the choices and *incentives* facing people other than the students themselves, and the rationality of their individual actions. This way we will understand better both the dynamics of current assessment systems, and how to improve them.

In an interview a few years ago, Mrs Thatcher (as she then was) made a still notorious remark. She informed the interviewer that 'there is no such thing as society'—a phrase which has been recycled endlessly ever since, invariably by those wishing to disagree with and attack it. Yet taken in context, what she said was actually perfectly true: one would say banal, except that we are surrounded by policy-makers who ignore it completely. What Mrs. Thatcher said was 'There is no such thing as society, only people.' Now obviously, at one level, that is indeed nonsense. We are social creatures, with socially imbued values: we need and exist in and through society and through social rights and obligations. But if she had said—and this was the context of her remark—'society doesn't do things, people do', then it would have been much harder to disagree with her and to miss the point she wanted to make.

To paraphrase: educational institutions don't administer tests, individual administrators and teachers do. It is not the teaching profession which teaches, it is teachers. And it is not the student body which learns, it is large numbers of individual students. Collective nouns have a seductive and misleading quality, because they imply action in unison. The verb is in the singular, *ergo* the action is the same. But why should all teachers behave in the same way? Why should all students? Why, indeed, should all civil servants either?

Many government officials and policymakers are astonishingly prone to the delusion that, if they design a policy and pass it down the line, it will somehow self-implement: people will obey the regulations and behave in exactly the way that the policymakers envisaged. Of course, this assumption can go spectacularly awry. A recent example is described in Chapter 8 on England and Wales: the total refusal of secondary English teachers, backed by their heads and governors, to administer government-mandated but seriously flawed tests of English to their 14-year-old pupils. More often, though, the 'distortions' are far more subtle and varied, though no less widespread in their effects.

It is surely much more reasonable to suppose that people will do as they are told and behave as they are 'meant' to behave to the degree that it seems worth their while to do so. And this is where we come back to economics—not macroeconomics and the contribution of education to some supposed international battle between countries for educational and economic supremacy—but to the idea of *individual rationality*.

Economics is sometimes attacked for the simplistic assumption that individuals are all profit-maximizers, whose behaviour can be predicted on the basis of which course of action is the most profitable. That is not strictly correct. Economics does deal with that part of social life where transactions can be measured in terms of money, profit and loss, and in that context the degree of financial profit to be made (long as well as short term) does indeed turn out to be a very effective predictor of how people behave. However, the basic assumption of the paradigm is rather different. As noted above, it is that individuals are rational in their actual behaviour if not necessarily in how they describe it to others! That is, people base their actions on what they perceive to be the most effective means to their individual goals.

The development of rational choice theories has made this clearer by widening the contexts to which this approach is applied. Among the important formal propositions are that:

- Individuals make decisions on the basis of their tastes and preferences.
- The more successful a given course of action has been in the past, the more likely someone is to repeat it.

- The more of something an individual has, the less interested he or she will be in yet more of it.
- The greater the demand for a good, the more 'valuable' it will be and the higher will be its price. The greater the supply, the less valuable it will be and the lower will be its price.
- People are rational in the sense that they can arrange the choices they are offered into a coherent order of preference, and then make consistent choices among them.

(Put more formally, this last proposition means that, in choosing between alternative actions, a person will choose that one for which, as perceived by him at the time, the value, V, of the result, multiplied by the probability, p, of getting the result, is the greater. Note that this does not require the individual to know the precise, actual probabilities or engage in abstruse calculation: merely to use the available information in the best way possible.)

It is important to emphasize that the model of rational behaviour being advanced does not presuppose any particular sort of values or preferences. It is perfectly rational to be altruistic in one's choices, or be motivated by, for example, the 'intrinsic' rewards of learning rather than 'extrinsic' factors such as labour market success. What it does emphasize is that, in a world of scarce resources, people are, necessarily, constantly engaged in weighing up alternative means to alternative ends and choosing between them. If one knows what people's preferred ends are, as individuals, one can predict and explain a great deal not just about particular individuals' actions but also about the cumulative effects at institutional level.

It is important also to emphasize that the 'rationality' paradigm deals with how people actually behave, rather than how they say or think they behave—because the two are often very different. The model also accords extremely well with empirical evidence, and not just tautologically (see e.g. Heath, 1976; McLean, 1987). People do behave rationally—in terms of their own individual preferences, and the choices that face them. Moreover, while there will be many people, in any given situation, who have similar preferences, and face similar choices, the likelihood that these will all be the same, or all in line with what policymakers like to think they are, is slight indeed.

Political scientists have made considerable use of this approach to look at the behaviour of bureaucrats, politicians and interest groups. For example, the way politicians in the US Congress 'trade' votes; the success of single-interest pressure groups; and the degree to which civil servants operate in pursuit of their own 'class' interest have all been analysed successfully in this way (McLean, 1987).

The world of assessment policy is also one in which developments are quite visibly the result of self-interested actions quite removed from the ideals and objectives of policy-makers. In Sri Lanka, for example, as both Chapter 1 and Chapter 13 make clear, assessment and certification policy have been substantially affected and moulded by the conflicting political interests of different ethnic and party groups. Chapter 1 also brings out very clearly one of the distinctive features of education in modern society: that it is now the dominant route world wide for passing on élite status, as well as for determining life-chances and mobility more generally. As such it is the focus for one of the most powerful and important of people's 'preferences', namely their desire to secure a good future for their own children. Family ties and family ambitions have a great deal to do with behaviour in the sphere of education and educational assessment, but this is all too often a taken-for-granted phenomenon whose implications are not fully thought through.

As Angela Little points out in Chapter 1, the fact that the British élite which created Sri Lankan education did not use the island's state schools for its own children, while the current élite does, is important in understanding educational assessment policy before and after independence. Sri Lankan parents and Egyptian parents effectively killed assessment reforms which they did not perceive as in their children's interests. (See Chapters 11 and 13.) At a more general level, the 'rational choice' perspective under discussion here would emphasize that senior civil servants and politicians everywhere are also likely to be parents. This fact, and the way the examination and assessment system works with respect to their own children, may be important factors in explaining national policy.

The area of educational assessment can also benefit from analysis of bureaucratic self-interest and institutional behaviour which has been developed in political science. The recent history of English National Curriculum testing, discussed in Chapter 8, provides an example. In twentieth century Britain, an action which breaks the law carries a considerable negative weighting—especially in the context of one's professional work. Governments can therefore count on teachers doing what they are told: but only up to a point. If the costs which obedience imposes become large enough, the possible benefits small enough, and if (equally important) the likely penalties for disobedience decline, one can quite quickly reach the point where, for most teachers, the incentive to disobey outweighs the incentive to do as they are told. Disobedience becomes rational.

In the case of England's Key Stage 3 tests for 14 year olds, this is precisely what happened. The costs of testing were great, the perceived benefits non-existent: and equally important, the more widespread resistance became, the lower the perceived risks were to any one disobedient individual. Equally important was the support of both head teachers and governors. Ironically, one of the same government's other reforms in recent years has increased the importance and autonomy of the Boards of Governors who are ultimately responsible for running English schools. The governors also objected to the Key Stage 3 testing: and, as non-civil servants, faced no major incentive to support a seriously flawed policy.

This English example underlines how unwise it is for governments to assume that whatever they order will necessarily happen. Yet, all over the world, it happens time and again. One's impression, during the lead up to the English 'Key Stage 3' débacle is that senior civil servants, and, even more, senior politicians, simply could not conceive of the possibility of disobedience. The rational choice perspective emphasizes that in fact there is always a choice: and that one cannot assume that rational individuals will always make the one that policy makers desire. If, when new policy was formulated, it was examined in terms of the incentives and concerns of those who must carry it out, there would be fewer policy disasters.

The remainder of this chapter uses a number of examples to illustrate how the 'rationality' paradigm might indeed improve our understanding of assessment systems, and of the gap between policy aspirations and the realities of implementation. It examines, in turn, the following questions:

- 'Payment by results', league tables and other accountability systems are meant to encourage teachers to be more effective. What incentive do they give pupils to change their behaviour?
- Advocates of school-based assessment by teachers (as compared to externally set tests) often argue for its more positive effects on pupil motivation and learning. Is the form of assessment really a key factor in determining pupils' motivation?
- Examining agencies are often required by governments to provide external validation and monitoring services to ensure consistency of standards. Can this be done in a context of competing educational institutions?

National testing programmes: high stakes for whom?

'Payment by results', league tables and other accountability systems are meant to encourage teachers to be more effective. What incentive do they give pupils to change their behaviour?

The first example is one with which we are relatively familiar. Programmes of this type are described, in for example, the chapters on England and Wales, Chile, Korea and the United States. They are assessment programmes involving the use of the same tests for all children in all schools—maybe every year, maybe only for some classes—and their analysis and collation on a school-by school basis as a way of monitoring the schools' performance. 'League tables' of the sort published by the English government, for example, provide a public ranking of schools; in this case on the basis of their students' success in public examinations (at age 16 and 18). 'Payment by results', in which the results are used directly in deciding funding allocations is relatively uncommon. However, it was used in Britain in the nineteenth century in funding elementary schools, and is used there today for some parts of further (post-compulsory) education. In China, where élite or 'key' schools receive considerably greater resources than ordinary schools, a key school with consistently poor results on public examinations may be downgraded in status (Wang, 1993).

If you read the comments of politicians, at least in England and the United States, you find a conviction that this sort of programme will improve children's achievement, by forcing teachers to work harder and more efficiently, and rewarding those who are successful. However, if you stop and think about the incentives a particular teacher actually faces, the situation is rather different.

First, such a teaching programme unquestionably does mean teaching to the test—which may itself be quite different from the overall curriculum. Striking differences in the sort of work which predominates in American as compared to English or French primary classrooms can be quite directly related to the frequent use in the United States of particular sorts of 'objective' tests, and to the frequency of school district or state-wide mandatory testing programmes. In England and Wales, Her Majesty's Inspectors have remarked on the high degree to which, three years into the testing programme for 7 year olds, teachers had started teaching to particular test items from previous years' assessments. Madaus and Greaney (1985) have provided a detailed analysis of the effects of the primary school-leaving certificate examination which was administered in Ireland between 1943 and 1967: and in Chapter 1 we have also cited studies which support the same point—that the content of previous examinations comes to define the classroom curriculum.

This is obvious enough, and not particularly unexpected; although advocates of large-scale testing programmes do tend to underplay the degree of backwash on the curriculum as a whole. (See e.g. Popham, 1978, 1984.) However, there are other less generally predicted effects of this sort of testing, which derive from the fact that, at the margin, teachers may not actually be able to improve pupils' rate of learning by much.

Having pupils who are very different in their prior achievement and ability to make rapid progress and who are in direct contact with any given teacher for quite limited amounts of time, means that teachers have a strong incentive to find other ways of raising the scores on tests. If you put yourself in the teacher's position, it is quite easy to see what is likely to occur. One way is to control which pupils actually take the test. Madaus and Greaney (1985) describe how Irish teachers limited promotion further down the school in order to control the failure rate at the examination.

During the 1980s, England moved to an integrated system of examinations at age 16, with all pupils taking GCSEs (General Certificate of Secondary Education) instead of choosing between the more difficult O levels (Ordinary level General Certificate of Education) and the easier CSE (Certificate of Secondary Education). One reason for the change was the tendency of schools to steer marginal pupils towards the CSE. They did so in order to improve their pass marks, even though at that time there was nothing so formal as current league table rankings: but the result was to preclude many CSE pupils from academic progression.

League tables have comparable effects. As noted above, they compare schools' results on the two sets of academic external examinations: GCSE (taken at the end of compulsory schooling) and A levels (the traditional university entry qualification). *The Times Educational Supplement* in England reported in 1994 that, faced with leagues tables comparing A level grades, some independent schools were demanding that marginal pupils enter as private candidates, rather than forming part of the school's A level entry list. Other schools are refusing to let academically weak pupils re-take GCSEs, on the grounds that they are unlikely to improve their grades, and so will not help the school's league table position (Wolf *et al.*, 1994).

It is always difficult to quantify the degree to which students are excluded from tests and examinations in order to improve results, though few teachers (in private) would deny that it occurs. It is not something that occurs via some wholesale public barring of groups from the test papers, nor something on which schools keep records. Rather, it is a grey area, in which individual teachers make little or no effort to chase up low-achieving pupils if they are absent, and senior staff become increasingly inclined to expel disruptive or low-performing students. In England, expulsions have increased steadily in the last few years as schools become more overtly competitive, and as their exam results become a staple item in the local and national press.

In the United States, teachers and school districts maximize their performance by aiming their teaching at the particular items and style of a particular standardized test (of which there is a range of nationally accepted variants available). This produced the phenomenon, whereby pupils' scores in each of the individual states, as shown by state-mandated testing programmes, improved at exactly the same time as nationally administered tests showed a continuing decline (Linn, Graue and Sanders, 1990a, b). Indeed, every state in the union managed to perform at above the supposed national average.

If tests are marked by teachers themselves, a comparable phenomenon emerges. Over time, average grades become consistently higher, and 'grade inflation' becomes endemic. Again, it is not that wholesale 'cheating' occurs. Instead, marginal candidates are given the 'benefit of the doubt' on the grounds that other teachers and schools are doing it anyway. This view was expressed quite forcibly in 1993 by teachers in England's lowest-scoring authority on the 7-year-old SATs (standardized assessment tasks), who were convinced they were paying for their professional honesty. Vocational awards in the UK already operate on a form of payment-by-results system: further education colleges and trainers receive more money for students on training schemes if they complete a formal vocational qualification than if they leave without one. Since the teachers/trainers are also, in these cases, the assessors, scepticism about the standards applied and the value of these awards is widespread, and has been expressed openly by national television programmes (Smithers, 1993).

Although it is very difficult to discern and record exactly what is happening at the individual level, the cumulative results of 'grade inflation' over a period of time can be very striking. In the United States, where grades are awarded by the academic who has taught the course, for example, university 'Grade Point Averages' displayed a huge rise during the 1960s. As described in Chapter 1, doubts by parents regarding the objectivity of marking led to the rapid demise of Sri Lanka's experiment with teacher assessment..

These effects occur without any discernible benefit to overall student achievement. In the nineteenth century, payment by results to state-supported English elementary schools had a number of clearly observable effects, notably the triumphant nationalization of the whole system, as state inspectors exerted their authority over the voluntary-aided sector, and a perceptible narrowing of the curriculum (Arnold, 1908). It did not, as far as anyone can tell, have any discernible effect on standards of education nationally.

At this point I would like to offer a proposition: *the higher the stakes for the teacher, and the lower the stakes for the pupil, the less reliable the test results will be.* For example, large-scale national testing programmes of the type being implemented or discussed in so many of the countries covered by this book are often described as 'high stakes'. This is actually often a misnomer. They may well not be high stakes for the pupils at all—only for the schools and the teachers. Therefore there is no reason for the pupils to make enormous efforts to do well, or be spurred on by the tests, and little potential for the tests to improve the quality of student learning.

A rational-choice perspective on assessment policy underlines the need to think about the different choices and motivations of the different 'players'. Too often, it seems, we ignore the key people in an assessment system; the ones actually being assessed. How much do they care about the results of a given assessment exercise? Do they have any particular reason to work hard, turn up, take the thing as seriously as their teachers might like? It is unlikely that an entire cohort of students will doodle on the papers, or turn the technology tasks into paper aeroplanes: but how much the assessment matters *will* affect students' behaviour at the margin. Even thin margins accumulate.

The final historical example in this context is again drawn from the United States, where policy makers seem to have an abiding tendency to conceptualize education and assessment solely in terms of the adults involved, and to see the pupils as passively responsive recipients. Twenty years or so ago, as part of the activist programme of the Office of Economic Opportunity and the Great Society programmes, the US Federal Government decided to experiment with 'performance contracting'. A number of optimistic contractors were convinced that what was wrong with learning and education was the lack of incentives to teach and learn, and therefore undertook to 'deliver' higher achievement by contracting with schools on a payment by results basis. The experiments were extremely carefully monitored, with outside administration of pre- and post-tests, using independent researchers. The schools concerned were elementary schools in largely deprived areas.

The experiment was a complete failure (Gramlich and Koshel, 1975). The schools in the sample did not show any improvement in performance. The pupils did not progress at any speed over and above what might have been expected. The teachers and the contractors did not win their bonuses and the programme was ended. Washington-based policymakers grew just a little bit more cynical about the possibility of any policy delivering sustained improvement in the schools.

Yet there was no good reason to expect that the policy would succeed. The teachers involved in the programme might have a reason to change their behaviour, assuming they had any idea of how to do so in a productive way. But the students? They took the tests, but no-one was paying them. Nothing rode on the results. Why should they change their behaviour? Work harder? Learn more? They didn't.

Pupil motivation

Advocates of school-based assessment by teachers (as compared to externally set tests) often argue for its more positive effects on pupil motivation and

learning. Is the form of assessment really a key factor in determining pupils'
motivation?

Formal mass education has had rather a short history, in the scale of human development: but it has nonetheless been around for long enough for a few quite radical experiments to have been tried. One rather depressing conclusion has, as a result, been quite well established. It is that, for most students, in most educational systems, 'intrinsic motivation' to learn (out of interest in the subject, a desire to know more or improve intellectual skills) is rather weak. Students work, for the most part, because and to the degree that success at school is necessary in order to have any chance of success in adult life, and especially to the degree that school results determine progress to more selective institutions which in turn promise better life-chances, or to obtain certificates and qualifications which license them to pursue desirable careers. As Lewin *et al.* (1994) point out, the most ambitious recent attempt to replace extrinsic with intrinsic motivation—the Chinese reforms of the Cultural Revolution which abolished all forms of academic examination and selection—failed totally. It resulted in the Chinese student population losing motivation altogether rather than in the development of intrinsic motivation in place of the old incentives.

Given that intrinsic motivation is relatively rare, and difficult to develop via top-down directives, the rational choice perspective allows one to make quite confident statements about the implications for education. Recent Chinese experience provides a particularly dramatic example of reforms which have foundered because of a misconstrual of how and why students learn, but European policy debates provide comparable scenarios.

In the last few years, policymakers in England have become increasingly interested in drawing comparisons with the country's major European neighbours, especially at upper secondary levels of education. This interest has been fuelled by a conviction that other countries' systems work better than England's; a view which is now treated as a given by most journalists and politicians. In fact, there is very little clear evidence on the relative attainments of, say, English, German and French 18 year olds. However detailed comparative studies do generally indicate apparently greater degrees of motivation among French and German students in *non-university bound* classes than among comparable English ones (see e.g. Prais, 1995; Steedman, 1992; Wolf, 1992). Motivation in this case is used not to refer to any inner feelings of interest and commitment, but simply as a description of what Americans call 'time on task'—that is, regular attendance in class, behaviour in class, work completed.

There is no indication, in any of the studies, that teaching methods are dramatically different in say, Germany and England, although the former is generally more traditional. What, then, is the explanation for these differences in behaviour?

The simple answer is that less academic students in Germany (and France) tend to work harder (be more motivated) than their English peers because the penalty for failure in school is far greater. The fuller and more complicated answer involves analysis of the whole system of incentives facing older students in a given society and underlines the way in which educational institutions are intertwined with specific features of a society's labour market and social structure.

German education has two distinctive features which are often cited admiringly by other countries' educators. The first is that school assessment does not, at any point, involve externally set and marked examinations. Everything is the responsibility of the class teacher: and the generally high status of German teachers is linked to this arrangement. The second is the ability of the 'Dual System' of apprenticeship, so called because it combines workplace and classroom teaching, to absorb virtually all school leavers who are not university bound. German youth unemployment is, in the 1990s, the lowest in Europe: and success rates for young apprentices are also extremely high. Almost all those entering will complete their award successfully.

The German system is indeed impressive. However, it is worth exploring just how this system creates a hard-working highly conformist school population, with proportionately the smallest group of unqualified 'drop-outs' in Europe.

First of all, the fact that assessment is by the teacher does not mean that it is in any sense 'low-stakes' from the pupil's point of view, or low in the stress it creates for students. Students who do not complete a year success-fully—that is students who fail on their teacher assessments—are obliged to repeat the class. Moreover, results on class assessments are crucial in determining students' progress within the academic sector, and in deter-mining what sort of apprenticeship a school leaver obtains, and where.

Not all trades, and not all firms, are equal. It matters very much what sort of apprenticeship you do, *and where you serve it*. An apprenticeship with Siemens or BMW is in no way the same as one with a small trader (Soskice, 1993a, 1993b). Many skilled workers are to be found in occupations other than the one they trained for; and the relative prestige of their qualification is therefore important beyond its immediate remit. Moreover, Germany is a country in which the 'shelf-life' of qualifications is extremely long. Life chances well into middle-age are significantly related to one's initial qualification: and you cannot turn around and enter a good apprenticeship in your twenties having failed or dropped out a decade earlier. Teenagers are thus involved in serious competition for good apprenticeship places; and non-university bound students are very aware that their school marks will nonetheless have a huge effect on their future lives.

France resembles Germany in the importance attached to formal quali-fications and the extremely poor prospects facing unqualified school leavers. If you drop out at 17, 19 or even 21 there are very few opportunities to climb back on to the education ladder. In England, by contrast, many of the qualifications open to young people in the lower half of the achievement distribution seem to have little effect on their life chances and future earn-ings (Bennett *et al.*, 1992). On the other hand, it is much easier in Britain than in most European countries to come back into further and higher education as an adult (Wolf and Rapiau, 1993). In the United States, second and third chances are even more easily obtained. Little wonder, then, that in England and the United States researchers tend to report more unmotivated pupils, more classroom disruption and less 'time on task'.

Focusing on the incentives which students face helps explain differences in motivation, and also brings out two general lessons. The first is that the effect of assessment systems on students' behaviour and motivation must be understood in terms of the whole structure of the labour market, not simply assessment methods *per se*, or even the structure of the school–university transition. The second is that there are few, if any, 'general' relationships between either the type of assessment used and student motivation; or between assessment methods and the sorts of control mechanisms which characterize an education system.

Prisoner's dilemma: barriers to standards in a competitive system

Examining agencies are often required by governments to provide external validation and monitoring services to ensure consistency of standards. Can this be done in a context of competing educational institutions?

The final example offered here is a simple application of game theory. Game theory is an approach based on the principles of 'rationality' and one which has been developed considerably by microeconomists (three of whom won the 1994 Nobel prize for economics for their game theory), by political scientists interested in bureaucratic and institutional behaviour and by biologists interested in evolution.

'Games' are stylized ways of representing recurring situations in which 'players' manoeuvre for advantage. The best known game of all is the 'prisoner's dilemma'. It is illustrated in Fig. 17.1 and will already be familiar to many readers. The 'prisoner's dilemma' is so well known partly because, in the original one-off, or one-occasion, formulation, the inevitable outcome

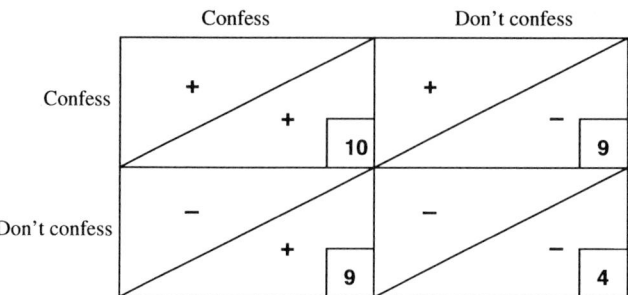

Figure 17.1 The prisoner's dilemma.

is so dramatically the worst one possible for both players, but also because the parallels with many social situations can be grasped very easily.

In the classic version of the game, two prisoners have committed a crime together. They are both under arrest and unable to communicate with each other. In order to force a confession, the authorities offer *each* prisoner, *separately*, the following deal:

- If you confess and your companion does not, he will get nine years, and you will be let off scot-free.
- If you both confess, you will get five years each.
- If neither of you confesses, you will get two years each.

Figure 17.1 summarizes the choices facing each prisoner. The four quadrants show the consequences of given actions for each prisoner separately and the *combined number of years in prison which will follow*. The box in the upper left corner shows the consequences if both confess. It is clearly the third worst choice for each individual—and the worst of all in terms of the *total* number of years' imprisonment which it implies. Yet this is, quite inevitably, the choice that their dilemma will lead them to make. Why?—because each prisoner will, in isolation, reason this way:

- If I don't confess, and the other guy doesn't either, we will only get two years each.
- But it is also quite possible that I'll keep quiet and then find the other guy has confessed. I'll then end up with nine years in prison. So not confessing is really risky.
- On the other hand, if I do confess, I may strike lucky—he may stay silent, and I'll get off completely. At worst, I'll get five years, which is better than nine. I'll confess.

The paradox which makes the prisoner's dilemma so intriguing is that both participants end up defecting even though they both know that they would be better off cooperating. A number of similar scenarios can be found in education: including that referred to here, the maintenance of consistent standards in a competitive environment.

The dynamic of the 'game' is very apparent in the UK because of quite dramatic and recent shifts in the way post-compulsory education is organized and funded. Over the last few years, publicly funded educational institutions throughout the UK, but especially those involved in 'further' (post-compulsory, non-degree level) and higher education, have increasingly been enabled and encouraged to compete with each other for students and for funds. Funding is tied to student numbers, and there is strong encouragement to expand numbers. However, there is also overt competition among further education institutions for funded 'training contracts', providing training and education to young and unemployed people on training schemes. This gives further education colleges a strong incentive to keep costs down, and also militates against inter-institutional cooperations. This incentive is strongest for college principals, and weakest for teachers, whose salaries are fixed through national pay bargaining, and to whom collaboration and contact offer many immediate and valuable benefits—a difference which itself creates tensions.

An additional pressure has recently been introduced in the form of 'payment by results'. Government training contracts provide higher payments to institutions for those students who complete their courses successfully and obtain a qualification than for those who fail.

These pressures operate in a system where almost all assessment is carried out not through external examinations but by the course teachers or tutors: either alone or as part of an internal team. (This degree of reliance on internal assessment is unusual. Other European countries have at least some external assessment for vocational awards.) Quality control is supposed to come from 'internal verification' of standards. For example, in the National Vocational Qualification system which now covers the bulk of awards for 16–18 year olds, each institution is supposed to appoint an internal verifier whose job it is to ensure that national standards are maintained. An external verifier visits once or twice a year for a day but is largely concerned with the processes being followed rather than with direct comparability of standards between institutions (Wolf *et al.*, 1994).

It is quite inevitable, in such a system, that the pressure is to pass rather than fail marginal students. It is also quite inevitable that standards between institutions will tend to diverge, with the 'norm' becoming what the average local rather than the average national student achieves. Only regular networking among assessors can maintain common criteria of judgement (Black *et al.*, 1989; Wolf, 1994).

Among students deciding where, among competing institutions, to enrol, there may be short-term advantages to choosing an 'easy' option. In the longer term, however, as this becomes known, their qualification will become devalued. This process is familiar to anyone who knows higher education in the United States, and is also similar to the way the German apprenticeship system works. Many players in the current British system are also well aware of the danger that their 'product' will become devalued and lose credibility, just as they are anxious about the way in which they or their institution fails to follow all the relevant rules. They are, however, caught in a classic variant of the prisoner's dilemma, illustrated in Fig. 17.2.

In the long term, it would be better for everyone to be in the top-left corner: for educational institutions (and the award-giving bodies who oversee assessment and verification) to express their anxieties, confess where things are not working, and so force change on the system. However, no one institution or player feels able to do this. Because they are involved in all-out market competition—and again this applies to both the colleges and the awarding bodies—they feel that, if they speak out, the others will close ranks against them, and deny that they suffer from this problem at all. Similarly, if they devote themselves to maintaining quality, they will simply lose out on bids and student through-put to their corner-cutting competitors. We therefore have the classic situation in which no individual dares speak out, even though this is clearly the 'best' situation for all: and we instead arrive in the lower right-hand corner, with a system operating well below optimum.

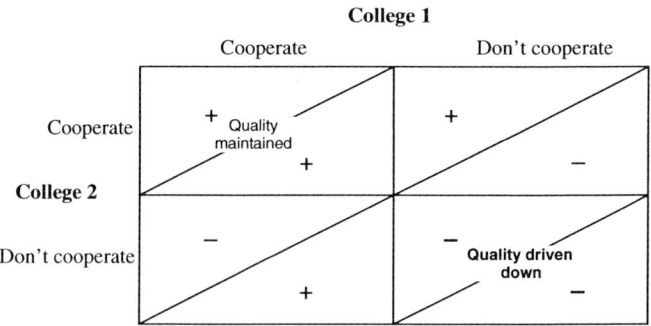

Figure 17.2 Cooperation between colleges in a competitive sector as a case of prisoner's dilemma.

Conclusion

A major theme of this book has been that of tension: notably that between assessment for monitoring, assessment for selection, and assessment that actually promotes learning. This chapter underlines the need to see this tension, and these different purposes for assessment, as interacting with other, related ones which are from the varying interests of different actors in the assessment system, and the different incentives (positive and negative) which they face.

A number of authors, notably Gipps and Somerset (Chapters 15 and 16), have emphasized that we do not have to be wholly pessimistic about the possibility of reconciling different assessment objectives. This is true. Different contexts and different sets of incentives produce different outcomes, and these can be positive and beneficial. However, what the rational choice perspective underlines is that no solutions are likely to be context-free and universal. The conclusions of this chapter therefore complement the arguments advanced by Angela Little in Chapter 1, where she argues for the importance of understanding assessment in the context of national (and regional) systems and cultures which will have unique and unre-producible features. The rational choice perspective in turn emphasizes the need to analyse the individual incentives which teachers, administrators, assessors and students face, and the way in which these are related to features of particular societies, and especially those of their labour market. Only if these are well understood in advance, can assessment policy hope to achieve the ambitious objectives which are now set for it world-wide.

References

Arnold, M. (1908). *Reports on Elementary Schools*, 1852–1882 (new edition).

Bennett, R., Glennester, H. and Nevison, D. (1992). Investing in Skill: To Stay on or Not to Stay on. *Oxford Review of Economic Policy*, **8**(2).

Black, J. H., Hall, J., Martin, S. and Yates, J. (1989). *The Quality of Assessments: Case-studies in the National Certificate*. Edinburgh: Scottish Council for Research in Education.

Gramlich, E. and Koshel, P. (1975). *Educational Performance Contracting: An Evaluation of an Experiment*. Washington, DC: Brookings Institution.

Heath, A. (1976) *Rational Choice and Social Exchange*. Cambridge: Cambridge University Press.

Lewin, K. M., Little, A. W., Xu Hui and Zheng Ji Wei (1994). *Educational Innovation in China: Tracing the Impact of the 1985 Reforms*. London: Longmans.

Linn, R. L., Graue, M. E. and Sanders, N. M. (1990a). *Comparing State and District Test Results to National Norms: Interpretations of Scoring Above the National Average*. CSE Technical Report 308. UCLA Center for Research on Evaluation, Standards and Student Testing.

Linn, R. L., Graue, M. E. and Sanders, N. M. (1990b). Comparing state and district test results to national norms: the validity of the claims that 'everyone is above average'. *Educational Measurement: Issues and Practice*, **9**(3), 5–14.

Madaus, G. and Greaney, V. (1985). The Irish experience in competency testing: implications for American experience. *American Journal of Education*.

McLean, I. (1987). *Public Choice: An Introduction*. Oxford: Basil Blackwell.

Newman, J. H. (1852). *Discourses on University Education*.

Popham, W. J. (1978). *Criterion-referenced Measurement*. Englewood Cliffs, NJ: Prentice-Hall.

Popham, W. J. (1984). Specifying the domain of content or behaviours. In Berk, R. A. (ed.) *A Guide to Criterion-referenced Test Construction*. Baltimore, MD: Johns Hopkins University Press.

Prais, S. (1995). *Productivity, Education and Training*. Vol. II. London: National Institute for Economic and Social Research.

Smithers, A. (1993). *Dispatches* report on education. All our futures: Britain's education revolution. London: Channel Four Television.

Soskice, D. W. (1993a). The German training system: reconciling markets and institutions. In Lynch, L. (ed.) *International Comparisons of Private-sector Training*. NBER Labor Market Series. Chicago, IL: University of Chicago Press.

Soskice, D. W. (1993b). Social skills from mass higher education: rethinking the company-based initial training paradigm. *Oxford Review of Economic Policy*, **9**(3), 101–113.

Steedman, H. (1992). *Mathematics in Vocational Youth Training for the Building Trades in Britain, France and Germany*. Discussion Paper No. 9. London: National Institute for Economic and Social Research.

Wang Binhua (1993). Comparing CESE in the People's Republic of China and GCSE in the UK. In *Educational Assessment: Sino-British Perspectives*, Little, A., Wang Gang and Wolf, A. (Eds). London: ICRA, Institute of Education.

Wolf, A. and Rapiau, M.-T. (1993). The academic achievement of craft apprentices in France and England: contrasting systems and common dilemmas. *Comparative Education*, **29**(1).

Wolf, A., Burgess, R., Stott, H. and Veasey, J. (1994). *GNVQ Assessment Review Project: Final Report*. Technical Report No 23, R&D Series, Employment Department Learning Methods Branch.

Wolf, A., Scharaschkin, A. and Pettitt, A. (1994). *GNVQs 1993–4: A National Survey Report*. The interim report of a joint project: The Evolution of GNVQs: Enrolment and delivery patterns and their policy implications. London: Further Education Unit.

Wolf, A. (1992). Mathematics for vocational students in France and England: contrasting provision and consequences. National Institute Discussion Paper 23.

Afterword

The theme of this volume has been the new and expanding demands placed on educational assessment. Assessment for selection has now been with us, in most countries, for centuries rather than decades. Assessment for certification also has ancient roots, although, rather surprisingly, it has not been a major characteristic of some of the world's larger systems. It too is becoming increasingly, rather than decreasingly, important. To these, now, are added assessment for the promotion of learning, and assessment for the monitoring and management of education systems: the former espoused by international and national policymakers and teachers alike, the latter the focus of tensions between government funders and those responsible for delivering education.

As noted in the Foreword, every chapter here attests to the growing demands placed on assessment systems, and the desire to combine all these objectives. The country case studies also make clear what tensions this can create. This is not a question of wealth and resources: England and Wales and the United States are the richest of the countries profiled here. In both, classroom teachers have experienced directly the tensions created by assessment reforms intended, by politicians, to create tighter control and higher standards at one and the same time. In Sri Lanka, Egypt and South Africa, educational assessment policy has been visibly entangled with major political issues and struggles. Even in countries where recent reforms have been introduced without major upheavals—China, Chile, Korea—the political tensions attendant on a high-stakes activity are never far from the reformers' minds.

Can we draw any other, general conclusions, or lessons for the future? At first reading, diversity appears to be the norm. Harold Noah argues this strongly, noting that although colonized countries and some of the newly industrializing systems have either borrowed assessment systems consciously or had those of others imposed, in many respects 'national idiosyncrasy is, and has been, the rule' (p. 94). While assessment may be on the increase everywhere, no other country has experienced the explosive growth in standardized testing described for the United States. China has decentralized dramatically, allowing provinces and municipalities to develop their own secondary school graduation certificates, at the same time as South Africa is centrally and rationalizing its range of qualifications. While many countries are experimenting with new ways of monitoring performance, no-one has attempted anything which attempts to combine multiple assessment objectives in the fashion of the National Assessment programme described for England and Wales.

Nonetheless, a number of general conclusions can be drawn.

The first relates to the ambition of those involved in the creation of assessment systems. This is very clear in the discussions of international agencies' activities. Vinayagum Chinapah sees the UNESCO Monitoring project as a way of significantly upgrading the quality of basic education in countries whose combined populations run to thousands of millions. He is confident that the procedures adopted by the countries which first participated in the monitoring project can be handed on to other later entrants. Marlaine Lockheed describes the rapid growth of the World Bank's support for educational testing, associated with an expansion of objectives to encompass educational management and quality improvement alongside certification and selection. Benefits can already be observed, she argues; in the future we can expect an increasing, positive impact on children's learning throughout the developing world.

At the level of individual countries, one is also struck by the ambitions harboured for assessment's impact. In the United States, past disappointments apparently have not dented the enthusiasm of policy-makers for reform through assessment. In England and Wales, although the scale and objectives of national assessment have been scaled back, the system is still regarded as an important—perhaps the crucial—instrument for raising standards, and for making a whole system more 'accountable' to users. In Chile, the government sees its monitoring system as the key to transforming education, especially in the weaker schools and districts.

The countries most ambitious for what assessment can achieve are also those most likely to combine several objectives in one system. Some would judge such ambition overweening. Marlaine Lockheed, for example, argues that 'different test designs are needed for different test purposes' (p. 31). Nonetheless, some of our individual authors share the optimism that has underlain recent assessment reforms. Caroline Gipps argues that it is possible to combine the different objectives of assessment, and that selection and monitoring need not be at the expense of learning. Similarly Anthony Somerset emphasizes the virtuous circle that examinations can create. They can, he argues, 'become a spearhead of educational improvement rather than an obstacle to it' (p. 282).

Overall, however, the papers here offer more in the way of warnings than rejoicing. Whenever they survey reforms which have been in place for more than a short time, the message is one of reality falling far short of expectations. While assessment reform in Sri Lanka was unusual in having focused widespread political protest, Egypt's attempted reforms have been equally abortive. In England, ill-conceived national assessment reforms destroyed ministerial careers and created a wholesale and successful revolt across the whole school system. George Madaus recounts a whole series of assessment-based attempts at school reform, with one more initiative following hard on the heels of its failing predecessor. Harvey Goldstein emphasizes the limitations of what international assessment comparisons can achieve, and argues for a more realistic and limited appreciation of their potential. And Angela Little cautions that the type of 'multi-purpose' assessment system which Gipps outlines requires a demanding set of circumstances that very few countries can provide.

Another common theme is the continuing dominance of a 'rational planning' paradigm. Government policy-makers, in particular, seem, across the world, to have a vision of how assessment systems work which takes surprisingly little note of the social context in which education operates. In part this reflects the fact that education has been relatively little affected by market ideologies. Although there has been considerable rhetoric in some countries about markets in schooling, education remains overwhelmingly state-financed and state-run. The commitment to providing for all children's access and opportunities means, we believe, that this is unlikely to change.

However, this seems, less fortunately, to be combined with a tendency among policy-makers (described by Alison Wolf) to see their systems in idealized terms, operating as their creators would like them to, removed from cost and time constraints or from the social and competitive pressures affecting students and teachers. In the UK, a number of commentators have remarked on the irony of a government which pays lip-service to market principles (which carry with them all the 'messiness' of bottom-up, decentralized change) while centralizing education to a degree unique in British history. Centralized testing programmes tend to win support from those who believe in central management and control, and the power of large-scale 'management systems' based on careful analysis of quantitative data, and who also distrust local decision-makers and teachers. This philosophy is one reason why reforms tend to be so ambitious. The limitations of such central planning discussed by Hayek as the 'rationalistic fallacy' also explain why reformers are so often disappointed.

The third theme to emerge from this volume is of growing internationalization: especially in the world of ideas. This may seem to contradict our remarks about continuing diversity: but what comes through very clearly from all the chapters is the extent to which reformers in one country are affected by what they perceive to be happening elsewhere. This does not necessarily lead to convergence because, at any given moment, there are time-lags, local (often unanticipated) pressures which drive reforms off course and, of course, a range of different ideas competing in the international arena. However, it does mean that it is increasingly difficult to understand what is happening in a country without taking into account not only the pressures of an increasingly global economy but also the influence of an increasingly global community of discourse. In the case of developing countries, of course, the 'international' element is even clearer, as the influence of past colonial powers is succeeded by that of international agencies offering loans tied to particular reform remedies.

We would end on both a pessimistic and an optimistic note. Assessment started, for most people and countries, with high-stakes selection. Far from this being less important than in the past, the evidence suggests that it remains of equal or growing importance to students worldwide, albeit at later ages in the richer parts of the world. Angela Little, in the first chapter of this book, describes vividly the pressures on students in Sri Lanka, and how much more severe they can be, at a younger age, than in England. More generally, international comparisons underline differences in the severity of selection pressures and their impact on students' lives: greater in Korea or Egypt (Chapters 12 and 11), less in the United States, for all its explosion of test use (Chapter 6).

Nonetheless, worldwide the trend appears to be towards greater, not less, selective pressure. In England, when teachers compare the present with the past, what they perceive is selection becoming more severe, not less. In the 'league tables' which list public examination results, the most successful school in England is the North London Collegiate School for Girls. In 1995, its headmistress was to be found bemoaning the fact that 'pressure for high grades is unrelenting'. Students and teachers alike, she argued, feel unable to address anything not clearly on the syllabus and likely to figure in the final examinations. Echoing Gipps and Somerset in this volume, she could praise the content of the English syllabus and examinations which keep education 'civilised, wide and demanding': but her main point is that selective pressure had been less, is currently severe, and is getting worse. Countries around the world would echo this.

The optimistic point relates to our greater understanding of the systems in which we operate, and which subject students to these pressures. Alongside the burgeoning demands made on assessment we have a growing body of work which shows us what is possible, and which provides clear pointers for how to improve future practice. This book encapsulates the knowledge which can be generated through an international dialogue which takes note of both shared objectives and diversity of context.

A.L.
A.W.

Contributors

Fouad Abou-Hatab
Director of the National Centre for
    Examinations and Educational
    Evaluation
Ain Shams University
Heliopolis, Roxy
Cairo, Egypt

Patricia Broadfoot
Professor and Head of School
School of Education
University of Bristol
35 Berkeley Square
Bristol BS8 1JA, England

David Carroll
15 St Bernard's Crescent
Edinburgh EH4 1NR, Scotland

Vinayagum Chinapah
Joint UNESCO-UNICEF Monitoring
    Project
Basic Education Division
UNESCO
7 place de Fontenoy
75352 Paris, France

Caroline Gipps
Professor and Dean for Research
International Centre for Research on
    Assessment
Institute of Education
University of London
20 Bedford Way
London WC1H 0AL, England

Harvey Goldstein
Professor of Statistical Methods
International Centre for Research on
    Assessment
Institute of Education
University of London
20 Bedford Way
London WC1H 0AL, England

Meng Hong-wei
Deputy Director
Department of Educational
    Evaluation
The China National Institute for
Educational Research
Bei Huan Zhong Lu 46
Beijing 100088, China

Sunderi Kariyewasam
(former) Director-General
National Institute of Education
Maharagama
Sri Lanka

Chan-Jong Kim
Lecturer
Chongju National University of
    Education
135 Soogok-dong
Heungduhk-gu, Chongju-shi
Chungbuk
Seoul, Korea 361-150

Angela Little
Professor of Education in Developing
  Countries
International Centre for Research on
  Assessment
Institute of Education
University of London
20 Bedford Way
London WC1H 0AL, England

Marlaine Lockheed
Education & Employment Division
The World Bank
1818 H Street West
Washington, DC 20433, USA

Peliwe Lolwana
Independent Examinations Board
PO Box 875
Highlands North
Johannesburg, South Africa

George Madaus
Boisei Professor of Education
  and Public Policy
Boston College
Campion Hall
Chestnut Hill, MA 02167, USA

Harold Noah
Professor Emeritus
Teachers' College
Columbia University
New York
NY 10027
USA

Josefina Olivares
Coordinadora Nacional
Project SIMCE
San Camilo 262
Piso 8
Santiago, Chile

Anastasia Raczek
Boston College
Campion Hall
Chestnut Hill, MA 02167, USA

Anthony Somerset
88 Prince Edward's Road
Lewes BN7 1BH, England

Jahja Umar
Director, Testing Centre
Office of Educational and Cultural
  Research & Development
Ministry of Education and Culture
J1 Gunung Sahari Raya No. 4
Jaharta Pusat, Indonesia

Alison Wolf
Professor of Education
International Centre for Research on
  Assessment
Institute of Education
University of London
20 Bedford Way
London WC1H 0AL, England

Author index

Subject index